The Future of Quantum Physics

Hans Christian von Baeyer

Illustrations by Lili von Baeyer

Harvard University Press

Cambridge, Massachusetts
London, England
2016

First printing

Library of Congress Cataloging-in-Publication Data
Names: Von Baeyer, Hans Christian, author.
Title: QBism : the future of quantum physics / Hans Christian von Baeyer ;
    illustrations by Lili von Baeyer.
Description: Cambridge, Massachusetts : Harvard University Press, 2016. |
    Includes bibliographical references and index.
Identifiers: LCCN 2016007855 | ISBN 9780674504646 (alk. paper)
Subjects: LCSH: Quantum Bayesianism. | Quantum theory.
Classification: LCC QC174.17.Q29 V66 2016 | DDC 530.12—dc23
    LC record available at https://lccn.loc.gov/2016007855

For Barbara

Contents

Introduction   1

I. Quantum Mechanics   **7**

  1. How the Quantum Was Born   9

  2. Particles of Light   21

  3. Wave/Particle Duality   31

  4. The Wavefunction   41

  5. "The Most Beautiful Experiment in Physics"   52

  6. Then a Miracle Occurs   63

  7. Quantum Uncertainty   73

  8. The Simplest Wavefunction   82

II. Probability   **95**

  9. Troubles with Probability   97

  10. Probability according to the Reverend Bayes   113

III. Quantum Bayesianism   **129**

  11. QBism Made Explicit   131

  12. QBism Saves Schrödinger's Cat   138

  13. The Roots of QBism   144

  14. Quantum Weirdness in the Laboratory   156

  15. All Physics Is Local   170

  16. Belief and Certainty   177

Contents

IV. The QBist Worldview   **185**

  17. Physics and Human Experience   187

  18. Nature's Laws   196

  19. The Rock Kicks Back   202

  20. The Problem of the Now   211

  21. A Perfect Map?   219

  22. The Road Ahead   223

  Appendix: Four Older Interpretations of
    Quantum Mechanics   235

  Notes   241

  Acknowledgments   247

  Index   249

Introduction

I'm a quantum mechanic in retirement. After fifty years of teaching the subject in universities, operating its mathematical machinery in my research, and struggling to bring its message to the general public by way of lectures, essays, books, and television, quantum mechanics has left its mark on me. It colors the way I think about the universe.

But ever since high school, when I discovered the magical world of *quantum billiards* and *quantum jungles* in George Gamow's classic Mr. Tompkins stories, I have suffered from a nagging feeling of unease about quantum mechanics.[1] It works flawlessly and has never let me down—or anybody else for that matter. But even as I used it and taught it, at some deep level I knew that I didn't really get it. I felt as if I were merely going through the motions that the pioneers of the theory choreographed long ago. Like all physicists I am fluent in Newtonian physics, also known as *classical physics,* and when the occasion demands, I rattle off its decrees, chapter and verse, the way an evangelist quotes the Bible, but I was never able to attain that same feeling of familiarity with quantum mechanics. There is a strangeness about quantum mechanics that is rooted not in its mathematical complexity but in the paradoxes and enigmas that have bedeviled it from birth. One of the most famous of those conundrums is the story of Schrödinger's hapless cat, which according to quantum mechanics is supposed to be both alive and dead at the same time. Other mysteries include the claim that a quantum particle can seem to be in two places at once, that particles can behave like waves and waves like particles, and that information appears to be transmitted instantaneously. Collectively, these puzzles have been called *quantum weirdness.*

I was reduced to taking solace from Nobel laureate Richard Feynman. Although celebrated as one of the leading quantum theorists of the twentieth century, he complained that "nobody understands quantum mechanics," including himself! His anguished admission didn't provide much comfort, though.

And then the unexpected happened. Just as I had started to plan my retirement and had resigned myself to the melancholy conviction that I would never feel completely at ease with the quantum, I stumbled upon an article by Christopher Fuchs, an expert in the frontier field of quantum information theory. Although I didn't understand the paper very well, it looked promising. So, following the tradition of the scientific community, I invited him to give a lecture at my academic home, the College of William and Mary in Virginia. He accepted and thus I began to learn about a new interpretation of quantum mechanics that he had helped to create. For reasons I will explain in this book, it is called *Quantum Bayesianism,* punningly abbreviated to *QBism.* QBism doesn't meddle with the technical aspects of the theory that has served me so well all these years and that has led to the invention of so many devices, which in turn have spawned entire industries that continue to transform our lives. Instead, QBism reinterprets the fundamental terms of the theory and gives them new meaning.

As Chris and I became friends, he patiently taught me how QBism manages to dispel much of quantum weirdness. For a decade we met at conferences and workshops in exotic places like an old Swedish castle, a high-tech think tank in Canada, a Swiss mountaintop hotel, and a dreary auditorium in Paris—wherever physicists gathered to debate the pros and cons of QBism. Chris and I visited each other's homes and families, exchanged innumerable e-mails, and emptied many bottles of wine together. Comprehension gradually dawned on me.

QBism is radical and profound, but it isn't particularly difficult to understand. I was so slow to embrace it due to the success of conventional quantum mechanics, which, for all its strangeness, is so astonishingly good at explaining nature and making verifiable predictions. Along with my generation, I was educated in a tradition that has jokingly been called the "Shut up and calculate!" school of physics. We were taught to accept quantum mechanics as fact, to use it for the purpose of explaining experiments and designing gadgets, and not to worry about its deeper meaning. "Get used to it!" was a politer version of "Shut up and calculate!" We were encouraged to push our philosophical misgivings aside and to get on with solving practical problems. Such a mind-set takes time to overcome.

Our complacent attitude began to change in the new millennium with the maturing of quantum information theory, which revealed unsuspected powers of quantum mechanics. Those were harnessed in such cool applications as quantum cryptography (for creating unbreakable codes) and quantum computing (for solving unsolvable problems.) The former is already a commercial reality, while the latter is poised to become practical in the not-too-distant future. Spurred on by a rapid progress in technology, the physics community is beginning to take a fresh look at the real meaning of quantum mechanics. A young researcher who expresses interest in studying its foundations is no longer brushed off as a dreamer. Chris and his collaborators deserve credit for stimulating a fruitful new interest in examining received wisdom—for stirring a pot that had been simmering on the back burner for far too long.

As I watched the message of QBism spread slowly through the physics community, I decided the time had come to write this book for people without easy access to mathematical formulas and equations. About twenty-five years ago, in a book about the effect of spectacular new images of individual atoms on popular physics, I wrote with more hope than conviction: "The bond of understanding we are . . . establishing with the atom will endow it with deeper meaning, until one day a profound and simple idea will resolve the enigma of the quantum." Well, that day has not arrived yet, but I have no doubt that just as advances in microscopy made the atom more familiar to us in the twentieth century, the profound and simple essence of QBism will nudge us closer to understanding the quantum in the twenty-first.

The first section of this book, titled "Quantum Mechanics," introduces the conventional theory in nonmathematical terms. To convey an intuitive sense of its meaning, I rely on metaphors and analogies to familiar things and everyday experiences. A high school physics course helps understanding but is not required.

In the next section, "Probability," I turn to a comparison between the traditional "frequentist" interpretation of probability as taught in high school and the less familiar Bayesian probability that put the B in QBism. Central to this discussion is the fundamental—and often neglected—distinction between formal mathematical probability theory and its real-world applications.

After this preparation, the heart of the book describes how quantum mechanics and Bayesian probability combine into Quantum Bayesianism and how this new interpretation dissolves quantum weirdness.

The last, somewhat more philosophical section, "The QBist Worldview," concerns the most significant lessons to be learned from QBism—its deeper meaning—the takeaway. QBism implies changes in the traditional attitude toward the underpinnings of the scientific view of the world. What is the nature of "the laws of nature"? Do those laws fully determine the evolution of the universe, or do we have free will to influence it? How do we relate to the material world, of which we are both parts and observers? What is time? Where are the limits of human understanding? Such questions, viewed from the QBist point of view, are touched on in this section. The final chapter takes a look at how QBism might develop from here on in.

QBism is more than old wine in a new bottle; more than yet another interpretation of quantum mechanics. Quantum mechanics has colored my view of the world—QBism has transformed it.

I

Quantum Mechanics

1

How the Quantum Was Born

A ccording to its inventor, the German physicist Max Planck (1858–1947), the creation of the quantum was an "act of desperation."[1] Spurred on by the technical challenges of converting public and private lighting from gas to electricity around the year 1900, physicists were exploring how glowing matter shines—how it emits light. When a hot object glows, be it a gas flame, the metallic coil of an incandescent lightbulb, or the sun, it radiates in different colors. By 1900 light was known to be some kind of wave, though it was not yet clear just what it was that was waving. Light waves, like water waves and sound waves, are described by their amplitude, or wave height, and their frequency, meaning the number of complete cycles, from crest to trough to crest again, that a stationary observer can record in one second.[2] We cannot see those cycles with the naked eye, but we know that light rays of different colors are distinguished by their frequency. Red light corresponds to slow oscillation or low frequency, yellow has an intermediate frequency, and blue light is characterized by high frequency—rapid jiggling. (A mnemonic: To recall whether red means slow or fast vibration, remember that the frequencies below those of the rainbow are called *infrared.* The prefix *infra-,* as in *infrastructure,* signifies below. Above the high end of the rainbow spectrum, we find ultraviolet light, with its prefix *ultra-,* for beyond.) In cases in which many colors are mixed together, as they usually are in nature, physicists ask: What is the relationship between intensity and frequency? In plain English: How much red light is emitted? How much yellow? How much blue? And so on through the rainbow.

In Planck's time experimentalists competed to measure the most exquisitely precise graphs of this relationship under ideal laboratory conditions. Plotting frequency along the horizontal axis and energy density, or brightness, along the vertical, such a "radiation curve" looks like a hill. The brightest colors emitted determine where the hill peaks. The radiation curve of the sun, for example, peaks in the yellow part of the spectrum. At the left, where infrared and red are recorded, not much energy is emitted. Edging toward higher frequencies, the curve rises to a maximum at yellow and then drops back down as the intensity diminishes at the high frequencies of blue, violet, and invisible ultraviolet light.

Theoreticians scrambled to explain these radiation curves by deriving them from the basic principles of physics. For years Planck worked on the problem with only partial success. Finally, in the waning

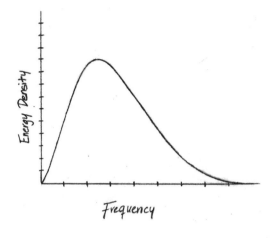

months of the nineteenth century, he tried using a statistical approach, which until then he had disdained. Hill-shaped curves are common in the field of probability and statistics. Consider, for example, throwing a pair of dice many times over and plotting the number of times you throw snake eyes, threes, fours, and so on up to twelve. Along the horizontal axis plot the *values* of the throws (the total number of pips showing on the two dice), ranging from 2 to 12, and along the vertical the *number* of times each value comes up. You are sure to end up with a hill—not perfectly symmetrical but low at both ends and rising to a maximum in the middle, at 7. The explanation of that shape is based on the idea of the *number of ways* a given throw can be realized. There is only one way to get snake eyes (1, 1) and one way to get a twelve (6, 6). But

11

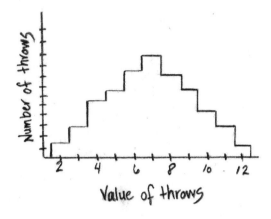

Value of throws

a seven can be obtained in no fewer than six different ways: (1, 6), (6, 1), (2, 5), (5, 2), (3, 4), and (4, 3). The intermediate values 3, 4, 5, and 6, as well as 8, 9, 10, and 11, can each be obtained in fewer than six ways. Since all combinations are equally likely to come up, the throw with the highest number of ways (the seven) wins, so the central peak of your graph, at seven, is neatly explained.

Planck set out to do something similar for the radiation curve. In order to do that, he needed to convert a continuous problem into a discrete one. Both the horizontal and vertical axes in the experiment with dice refer to countable quantities—both are measured by simple integers. In the radiation curve, on the other hand, the frequencies of light are measured by real numbers from zero to infinity. (The rainbow does not consist of the colors red, orange, yellow, green, blue, indigo, and violet recalled by the mnemonic Roy G. Biv but of an infinite, uncountable number of hues.) The vertical axis of the radiation curve is just as troublesome. The energy that a glowing object emits is also measurable but not countable. If he wanted to "count the ways," Planck had to approximate the smooth radiation curve with a graph with stepped sides—like a Mexican pyramid. If he made the steps small enough, they would be too tiny to perceive, and the jagged outline could stand in for the actual smooth curve.

Although Planck, along with some of his contemporaries, didn't believe in the reality of atoms, he had a good imagination. He knew that the heat energy of a glowing object is an expression of some kind of invisible motion. What we perceive as heat is really the imperceptible, internal jiggling or vibration of the material of the object. (You can turn motion into heat by simply rubbing your hands or by drilling into a hard solid with an electric drill!) With this understanding Planck invented an ingenious model that made both the frequencies and the energy countable.

The simplest device that stores energy and jiggles with a definite frequency is a *harmonic oscillator*. (The charming word *harmonic* stems from the role of oscillations in producing musical sounds.) An example of a harmonic oscillator, or oscillator for short, is a weight resting on a frictionless surface at the end of a spring that is itself attached to a wall. Other examples include

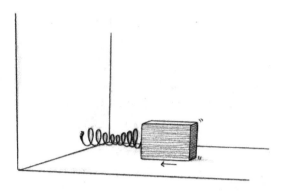

tuning forks, musical instruments, and pendulums. When at rest with its spring relaxed, an oscillator possesses neither kinetic energy of motion nor potential energy stored in the stretched or compressed spring. But after you give it a little push, its energy sloshes smoothly from kinetic to potential and back again with a fixed frequency whose magnitude is symbolized by the letter f. If there were truly no friction, its total energy would remain constant, and the graceful harmonious motion would continue forever.

As a temporary expedient, a mere mathematical trick, Planck imagined the total heat energy of the glowing object (say, a little ball of glowing gas) to be distributed among a very large (but not infinite) number of tiny oscillators of unspecified design, whose only functions were to store energy by vibrating at a definite frequency and constantly emitting and absorbing light at that same frequency. They were not supposed to model any of the countless other properties of the gas—not its chemical composition, density, or electrical resistance, for example. Planck's model was far-fetched, but visionary.

Later it became clear that Planck's little imaginary gizmos are actually quite real—they are the vibrating atoms and molecules that make up the glowing ball and that do indeed emit and absorb light. (The rigid wall in the fictional model represents the great mass of the gas that surrounds each vibrating atom and keeps it more or less in place.) Atoms are numerous, to be sure, but their number in any real object is countable (in principle, though it's hard to do in practice) and finite. Planck's oscillators, on the other hand, were as he put it, "a purely formal assumption and I really did not give it much thought." The point of this leap of the imagination was to break up the range of frequencies into a finite sequence of discrete, countable values analogous to the eleven discrete *values* from 2 to 12 of the throws of your dice.

Next, Planck had to divide the vertical axis, representing radiated energy, or brightness, into discrete steps as well, to correspond to the *number* of times each value shows up on your dice. To this end he made the strange, totally unheard-of assumption that each of his oscillators could only store energy in small equal portions—atoms of energy, as it were or, as Planck himself called them, "energy elements." This was a more consequential hypothesis than merely subdividing the frequency axis. For each oscillator he divided the energy into equal bundles, admitting the possibility that the magnitude of the bundles might be different, depending on frequency. If the energy in that bundle is called e, an oscillator could store a total energy of 0, or e, or $2e$, or $3e$, and so on. Note that this sequence can't possibly go on to infinity because there is only so much energy available in the whole glowing ball, so a single oscillator can store the total amount available and no more. This subtle point, in the end, made a crucial difference in the calculation. It kept the accounting nice and finite instead of running off to infinity.

In order to make a prediction for a real experimental radiation curve, Planck had to figure out the actual value of e. How much energy is there in one of these little imaginary bundles? Guided by the knowledge that if the amplitude were kept constant an ordinary oscillator's energy would increase with frequency, Planck assumed that the amount of energy in a single bundle is proportional to the frequency (symbolized by f) of that oscillator. (The faster you wiggle, the greater your energy of motion.) Mathematically, this means that the fundamental bundle e is obtained by multiplying the frequency by a small adjustable constant he called h. (An adjustable constant, also called a *parameter*, is a number that is fine-tuned to fit the circumstances but then locked in.) In symbols

$$e = hf.$$

Mentally shuffling that astronomical number of energy packets stored in that vast collection of oscillators, Planck was able to count the *number of ways* the total energy can be distributed among the oscillators and to plot a curve of energy versus frequency for the whole ball of gas. Just as in the case of your dice, the left- and right-hand ends of the resulting curve turned out to be lower than the central peak. By fiddling with the magnitude of h and adjusting its value to fit the data, he reproduced the experimentally measured radiation curves with astonishing precision.

Although this achievement earned him a Nobel Prize, Planck hoped for years that his energy bundles were nothing but computational aids and that a more refined model would restore unbroken continuity. He could not simply ignore the constant h or make it disappear because it shows up in his final formula for the actual radiation curve measured in the laboratory, but he hoped that the little oscillators and their tiny energy bundles were mere artifacts—like luminous grid lines projected onto a sheet of paper as an aid to drawing that are switched off in the end.

But Planck was wrong on both counts. The oscillators, as I mentioned, turned out to be atoms and molecules. Energy packets, in turn, would in time be called *quanta* (the plural of *quantum,* Latin for amount), and the parameter h, now named *Planck's constant,* became the fundamental coin of the realm of quantum mechanics. Planck's desperate trick turned out to be the opening act of the birth of modern physics.

In Einstein's hands Planck's little formula $e = hf$ became what might be called the icon of quantum mechanics, just as his $E = mc^2$ became the icon of relativity theory. Of the two equations, the latter is the more famous, but $e = hf$ is just as powerful. Whereas the relationship between energy and mass is derived from the more fundamental principles of relativity, Planck's link between energy and frequency was an unexplained axiom of the early quantum theory. Today it is regarded as a consequence of quantum mechanics, which itself rests on more fundamental principles.

In metric units the modern value of h is given by[3]

$$h \approx 0.000\ 000\ 000\ 000\ 000\ 000\ 000\ 000\ 000$$
$$000\ 000\ 662\ 606\ 957\ \text{joule-seconds.}$$

The scientific custom of writing $h \approx 6.63 \times 10^{-34}$ joule-seconds is more convenient for sure, but writing out the entire parade of thirty-four 0s, representing as many factors of 10, is a visual reminder of the inaccessibility of the atomic world to our senses. Our direct experience ranges from a visible horizon of, say, one hundred kilometers, or $1.0 \times 10^{+5}$ meters, to the thickness of a fine human hair, ten millionths of a meter, or 1.0×10^{-5} meters. For anything outside that narrow interval of eleven factors of 10, we require mechanical help in the form of telescopes and microscopes. But even those don't come close to reaching the unimaginably tiny dimensions of Planck's calculation. The realm of the quantum was revealed by reason, not directly by our senses—or even our measuring instruments.

Because he disliked his own energy bundles so much, Planck missed the immense significance of his little formula. That insight was left to Albert Einstein, who, just five years later, promoted quanta from mathematically convenient fiction to measurable reality. Einstein set out to investigate whether the energy emitted as light retains its discrete character during propagation. A Bavarian by birth, he once put the question in colloquial terms: "Even though beer is always sold in pint bottles, it does not follow that beer consists of indivisible pint portions."[4] Where Planck had thought of such portions as residing in matter, Einstein proposed that light itself consists of bundles of energy, which he called quanta and which were later renamed *photons*.

The ancient Greek philosophers known as the *atomists* had proposed that matter consists of individual particles. *Electrons,* the uncuttable particles of electricity, had been discovered late in the nineteenth century. Einstein proposed that light, like matter and electricity, might upon close examination turn out to be grainy too.

2

Particles of Light

We don't know exactly how Einstein came up with his radical and enormously influential ideas, but he did leave a few clues. When asked, "What is thinking?" he replied that it doesn't start with words or equations. Instead, he suggested, it begins with "the free play of images," a process we might describe as daydreaming, or doodling, or allowing mental images to tumble over each other like bits of colored glass in a kaleidoscope. Even that, Einstein continued, is not yet thinking. But if some pattern among the playful images pops up repeatedly, it may suggest a fresh concept. And if, in the end, that concept is put into words or mathematical symbols—eureka!—an idea is born.

In his miracle year of 1905, in which he shocked his colleagues with his special theory of relativity, Einstein also pondered the mystery of the photoelectric effect. When light shines onto the surfaces of certain metallic plates, it liberates electrons by knocking them out of the metal. Since electrons are negative, their departure leaves the plate positively charged. When the effect was studied in detail, it presented two puzzles. As expected, the electrons emerged with a variety of energies—presumably, they bounce around inside the bulk metal and slow down on their erratic way out. But for light of one specific color, there always seemed to be a maximum electron energy; an energy threshold that no electron could exceed. You can increase the intensity of the light—drenching the metal plate in a flood of optical energy that knocks out electrons in torrents—but that doesn't increase their top speed or maximum energy. What is holding them back?

The other puzzle of the photoelectric effect cropped up when different metals, as well as different colors of light, were compared. For each metal there was a threshold frequency of light below which the effect stopped. In other words, if the frequency of light was too low—if the color of light was "too red"—no electrons were liberated, despite the intensity of the illuminating beam. What's wrong with light from the red end of the rainbow that it should fail to dislodge electrons from a metal?

Neither of these two observations—the maximum energy of electrons and the minimum frequency of light—makes sense in the classical scheme of things. That light consists of waves was demonstrated conclusively at the beginning of the nineteenth century. Subsequently, physicists learned to describe those waves as weak electric and magnetic fields that oscillate rapidly as they propagate through space at the speed of light. Thinking about electrons as pebbles on the beach and light as ocean waves crashing in on them and knocking them around, an image that Einstein might have considered when he started doodling, does not suggest any reasons for the peculiar details of the photoelectric effect. But under certain circumstances, electrons would be limited to a maximum speed. Encouraged by the successes of atomism, imagine that the incoming light waves actually consist of identical discrete chunks of some sort. These chunks can't be real atoms or molecules because we know that light is not made of matter. But if the imaginary chunks of light of one color all had the same energy and one of these chunks hit a single pebble head-on, the pebble could absorb all of the chunk's energy but no more. (Pool players know that a rolling ball hitting a stationary ball head-on will transfer to the target its entire energy—and no more.) In this image, there would be a maximum among the electron energies, just as observed.

At this point Einstein would have remembered Planck's tortured reasoning, five years earlier, which had forced him, albeit reluctantly, to adopt the hypothesis that matter emits light in bundles with energy $e = hf$. Although the photoelectric effect that Einstein considered and Planck's radiation curves of glowing matter are unrelated phenomena, they both deal with the nature of light at its most intimate. Only Einstein's wide-ranging play with images suggested that the two sets of experiments—one concerning absorbed light and the other, emitted light—might reveal a common pattern. His crucial step was to extend the atomic hypothesis from matter and electricity—where it had succeeded brilliantly—to include light as well. Call it a chunk or a bundle or a quantum, today the "atom" of light is called a photon, and it is the second truly elementary particle discovered, after the electron. It served as a model for many other elementary particles to come—most recently, the famous Higgs particle, discovered in 2012 after a search lasting half a century.

Einstein replaced the image of the pebbles pummeled by waves on the beach with that of a stream of photons impinging on a crowd of more or less stationary electrons stuck in a metal plate. Every now and then, a photon hits an electron and gives up its cargo of energy e—vanishing in the process like a snowflake melting in your palm. The electron then races off helter-skelter, bouncing off nearby atoms in a zigzag path, and finally leaves its prison. It starts off with energy e and may lose some of that along the way but, and this is the point, *it doesn't pick up any more than that.* Increasing the intensity of the incoming light increases the number of absorbed photons, but each one carries the same energy e. The maximum energy that individual affected electrons absorb

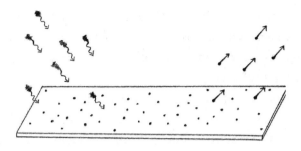

remains the same—only their number increases. That takes care of the first puzzle.

The solution of the second puzzle must have thrilled Einstein when he first glimpsed it. Why is there a minimum frequency—a "reddest color"—below which the photoelectric effect ceases to operate? The answer is that the electrical attraction of the positive nuclei of the metal holds the electrons in their plate— like frogs in a well. They can't escape unless a photon gives them a boost. And if that boost is insufficient, the electrons simply have to stay inside. If the color is too red, the frequency of the incoming light will be too low, and by Planck's formula the energy of each photon will be too feeble to supply the required boost. Each metal has a natural minimum frequency below which the incoming light, no matter how bright, cannot knock electrons out of the plate.

Proof of the validity of Einstein's model of the photoelectric effect, based on the image of photons hitting more or less stationary electrons, took more than a decade of careful experimental work, but the results were convincing when they came in: light consists of particles.

The experimental demonstration that light consists of waves is just as persuasive and much simpler. It was first achieved in 1803, about a century before Planck and Einstein's quantum hypothesis, by Thomas Young (1773–1829).

The unique signature of waves, which distinguishes them unequivocally from particles, is the fact that under special circumstances waves can cancel each other out, leaving nothing behind—a trick called *destructive interference,* which common sense informs us neither pool balls nor marbles nor any other ordinary particles can accomplish. Consider two identical waves arriving at the same spot from different directions. Where they meet they overlap in what is called a *superposition,* meaning they occupy the same position "on top of each other," like two superimposed photographic images. If the two waves happen to be perfectly out of step so that when a crest of one arrives a trough of the other meets it, the two will cancel each other for as long as they stay in sync. Such dark spots, where waves interfere destructively, are common in nature, if you know where to look. Ocean waves, sound waves, radio waves—even seismic waves and waves on jump ropes shaken by

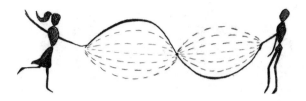

children—can develop such telltale dead spots. (If the two waves are in step so crest meets crest and trough meets trough, they reinforce each other in what's called *constructive interference*.)

The invention of the laser, itself a product of quantum mechanics, has made it easy to observe the destructive interference of light. YouTube lists videos of homemade demonstrations of light waves under the search term "Double-Slit Interference Experiment." One of them uses a laser pointer obstructed by a mask in the shape of a double slit made of small pieces of electrical tape on either side of a thin wire: ■|■. Shining the laser light through the two slits at a wall produces an interference pattern.[1] The beams from the two slits are perfectly in step as they leave the mask. However, at every spot on the wall light arrives from two different sources. Since the distances from the two slits to that spot are slightly different (except for a line down the middle), the waves will be in step or out of step depending on the exact location on the wall. What you see is a pattern of parallel lines on the wall, alternately bright and dark.

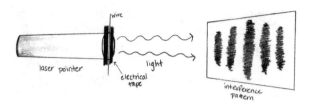

A quick aside is in order about the use of slits as sources of light rather than small openings such as pinholes. To make interference apparent, the pinholes must be small and close together. Because of this limitation, pinholes don't allow much light to pass through them. But if you substitute two thin slits, which can be as tall as you like, you get more light and a better image, even though the two sources remain both narrow and close together. For this reason the experiment is usually performed with slits instead of pinholes.

The bright lines on the screen are positioned where the light beams from the two slits reinforce each other. The dark lines, where the two beams happen to cancel each other, are proof that light consists of waves.

Actually, once you know that light consists of waves, you can find interference effects all over the place. Interference causes the shimmering colors in soap bubbles, for example. When a beam of light shines at the wall of a soap bubble, which consists of a thin film of water, it is reflected off two surfaces. The

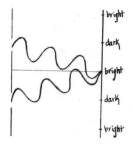

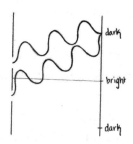

portion of the beam reflected off the inner surface is delayed a bit by passing through water and thrown out of step with the part reflected from the outer surface. The amount by which it is out of step depends on the thickness of the wall of water as well as the frequency, or color, of light. When the two beams recombine to reach our eyes, the portions of light that are out of step destroy each other and are removed from the spectrum, while the portions that happen to be perfectly in step are reinforced. Thus, different thicknesses of the bubble wall favor different colors, and those colors shift as the bubble twists, wobbles, and deforms. Nature, in her inimitably flamboyant way, reveals the waviness of light—almost as obviously as she shows us the waviness of the ocean's surface.

Other displays of interference include the rainbow colors reflected from CDs viewed obliquely, the iridescent colors of butterflies, the lovely hues of mother-of-pearl in seashells, the shimmering of oil slicks on asphalt in the rain, and even the patterns on the tails of peacocks. They are all nature's way of telling us about light waves. She is much more reluctant to reveal that light may also behave like a shower of pellets. It took an obscure phenomenon—the photoelectric effect—and the unique imagination of Albert Einstein to tease out that hidden aspect of the wondrous and ubiquitous stuff called light.

So how are we to think of light—as an electromagnetic wave speeding through space or as a stream of ghostly particles?

3

Wave/Particle Duality

Photons are strange beasts. If you were to repeat the double-slit experiment and preserve the images of the arriving photons (the way a paper target preserves the bullet holes made by a rifle), you could follow the gradual development of the image and observe both halves of the split personality of light, its wave/particle duality, at the same time. Turn the brightness of the light way down so that on average only one photon is emitted every minute or so. At first the screen is blank. Then a dot appears somewhere—*ping*—a tiny pinprick announces the arrival of a photon. After an interval of a minute or two, the next dot shows up elsewhere. The intervals between pings are random: *ping*—pause—*ping ping ping*—long pause—*ping ping*—short pause—*ping ping ping ping*. And so on. For a long time the dots appear to be scattered randomly over the screen. But after hundreds of hits, you begin to notice a pattern. Spaced at regular intervals, empty stripes cross the image, parallel to the orientation of the two slits. And if you wait long enough for thousands of photons to register, the characteristic striped pattern of double-slit interference emerges.

  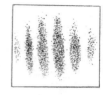

Discrete particles make the dots, yet the stripes provide irrefutable evidence of waves. You might be tempted to shrug your shoulders and point out that water waves also consist of myriad particles, namely, H_2O molecules. So what's so strange about light being both wave-like and grainy? The subtlety lies in the timing. Water waves (like fan waves in football stadiums) are made by countless units, each connected to its neighbors in some way, all acting in concert. But the photons from the laser arrive at long intervals, so there can be no possible connection or communication allowing them to coordinate their motions. They could have arrived hours apart instead of minutes, with the same result. It's as though ten thousand blind and deaf spectators in a stadium managed to perform a crisp fan wave—without touching each other. It smacks of magic. It's weird.

If the physicists of the early twentieth century found the wave/particle duality of photons perplexing, they were soon in for an even greater shock. Beginning in 1923, they learned not only that waves can behave like particles but that the reverse can happen too:

electrons, which had been thought of as particles, can act like waves. The proof of this astonishing claim proceeds in precise analogy to the double-slit experiment with laser light. The laser is replaced by a thin beam of electrons—variable in intensity like the laser—and the slits must be made much narrower and spaced far more closely than in the homemade light interference experiment. In place of a blank wall or a photographic plate, finally, a fluorescent screen flashes whenever an electron strikes it. But the result is exactly the same: dots appear at random intervals and unpredictable positions yet gradually build up a perfect pattern of parallel interference stripes. Read more about this in Chapter 5.

Wave/particle duality is embodied, with fine historical irony, in a father-son pair of British physicists who helped to lay the foundations for what eventually became known as the quantum theory. In 1906 J. J. Thomson (1856–1940), one of the grand masters of experimental physics in his time, won a Nobel Prize for proving that electrons are particles by tracing their parabolic trajectories through electric fields, mimicking the paths of golf balls flying through Earth's gravitational field. Thirty-one years later his son G. P. Thomson (1892–1975), following in his father's footsteps, was awarded his own Nobel Prize for proving that electrons are waves by demonstrating their destructive interference. The father, who was also a graceful writer, summed up the dilemma: "[The wave/particle view of physics] is like a struggle between a tiger and a shark: each is supreme in his own element but helpless in that of the other." Imagining a photon or an electron as a particle cannot explain double-slit interference for either one; thinking of them as waves explains nothing about the photoelectric effect or the arcing paths of electrons. The wave theory and the particle theory seem incompatible.

J. J. Thomson's quip refers to two theories, which differ from each other as fundamentally as a tiger and a shark, for describing both photons and electrons observed under different circumstances. That explanation doesn't satisfy our hunger for true understanding. The aim of physics is not merely to tell a convincing story about every object and every event in the material universe but to produce a single epic, a coherent theory for describing nature. No one was more driven by this passion for unification than Einstein, who had stirred up the rivalry between tigers and sharks in the first place. As early as 1909, four years after proposing particles of light and sixteen years before the birth of quantum mechanics, he predicted in a lecture at a meeting of German physicists: "I believe that the next phase in the development of theoretical physics will bring us a theory of light that can be considered a fusion of the [wave] and [particle] theories." He knew precisely what was needed, even though in the end he would not be entirely satisfied with the solution that was offered.

The trouble with wave/particle duality is easy to spot. Waves and particles are categories we have derived from observing the everyday, macroscopic, Newtonian world around us, and they are simply inadequate for the domain of the atom. Photons are not like ocean waves or bullets, nor are electrons. They both have certain properties in common with waves and particles, but they don't share their every characteristic. Why should they? We can't shrink, like Alice in Wonderland, to atomic dimensions to see for ourselves how elementary particles behave in their own environments. The best we can do is to use our imaginations to help us paint a consistent picture that accounts logically for all observations in our human-sized laboratories.

To mediate between the incompatible "wave" and "particle" categories, the term *wavicle* was suggested to describe the electron, but fortunately this ugly and uninformative word never caught on. More picturesquely, my late friend Rolf Winter, inspired by J. J. Thomson's animal analogy, likened the electron to a platypus. When explorers first brought platypus specimens from Australia in the eighteenth century, the learned naturalists at European universities declared them to be forgeries, stitched together from parts of other animals. "Mammals don't lay eggs," they said, and "reptiles don't suckle their young." "An animal that is both mammal and reptile cannot exist, so it is a hoax," they harrumphed. But the categories they had invented on the basis of their own limited observations turned out to be inadequate to describe the profusion of Earth's creatures. Similarly, photons and electrons are particles that can behave like waves and waves that can behave like particles. Like the platypus, they blithely ignore the categories we derive from inappropriate precedents.

To advance beyond the invention of useless new words like *wavicle* and beyond comparisons with exotic animals required a more radical approach. Einstein's 1909 call for a fusion of wave and particle theories wasn't answered until the birth of quantum mechanics in 1925, but that baby started kicking well before it saw the light of day.

In 1913 the Danish physicist Niels Bohr (1885–1962) constructed the first successful model of the interior of an atom. Conforming to the time-tested habit of physicists to start with the simple, he fixed his attention on hydrogen, the first and lightest element in the periodic table. Inspired by a bold analogy to the solar system, he described a lone electron circling the central nucleus the way Earth circles the sun. Only certain discrete orbits, whose radii are fixed by the magnitude of Planck's constant h, are allowed. Photons, with energies given by the Planck-Einstein

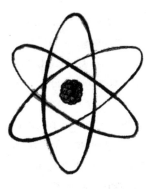

equation $e = hf$, are absorbed (or emitted) by the atom
as its electron leaps up (or down) the rungs of the
ladder of possible orbits. The resulting picture was
quickly refined to include elliptical as well as circular
orbits, to obey the rules of special relativity, and to describe more complicated atoms than hydrogen. Eventually, the celebrated "Bohr model" of the atom gave
rise to what became one of science's most recognizable
caricatures—the ubiquitous picture of an atom drawn
as a dot in the center of three ellipses representing the
trajectories of three electrons—presumably, those of
lithium, the third element in the periodic table.

This little icon, reproduced in countless variants,
is universally recognized as representing an atom
and has been adapted for use as the logo for high-tech
companies, government agencies, and consumer products. It flits across the screen between scenes in the
sitcom *The Big Bang Theory*, and throughout the world

it suggests power, be it of toothpaste or a think tank. Because the logo's message is so simple and convincing, it dominates high school teaching and defines the understanding of atomic structure for the majority of the public.

Unfortunately, it is also fundamentally flawed.

In 1919, just six years after introducing it, Bohr himself was forced to repudiate it because even then it badly misrepresented the prevailing understanding of how electrons behave inside atoms. The Bohr model describes the path of hydrogen's lone electron as an orbit around a hydrogen nucleus (also known as a proton). The resulting structure is as flat as a pancake—but we know by watching the atom interact with other particles that from the outside it resembles a fuzzy ball of cotton more than a pancake, at least in its normal, undisturbed state.

Worse, the logo suggests that the electron spends its life permanently separated from the nucleus by the radius of its orbit, a distance known as the *Bohr radius*. But experiments show that when it is detected in the atom, it can be found not only on the surface but anywhere *inside* the cotton ball.

The most egregious and unforgivable flaw of the Bohr model is more fundamental than these technical shortcomings. By assuming sharp, definite trajectories, the model ignores the wave/particle duality of the electron in favor of its particulate nature. The Bohr model is a throwback to Newtonian physics, in which a particle follows a precise, continuous trajectory and has a well-defined position as well as a definite velocity at each point of its path. Speaking about electrons in atoms the way we speak about the orbits of planets has been banished from the vocabulary of physics for a century.

By its ability to evoke a vivid mental picture, the Bohr model captured the imagination of popular science to an alarming degree. A monument to arrested development, it suggests that nothing has changed in atomic physics in a hundred years. No other fundamental science gives that impression: Not cosmology, with its breathless succession of new discoveries, such as the accelerating expansion of the universe and the enigmatic stuff called *dark matter* and *dark energy*. Not astronomy, with its daily output of stunning images of distant objects in luminous colors. Not biology, with its growing understanding of the structure of the brain, the subtleties of the human genome, and the mind-boggling products of evolution. The universal atomic icon is as dated as a drawing of a horse and buggy to represent a parking garage or a cartoon of the Wright brothers' airplane to point the way to an airport.

The Bohr model was an important step in the development of quantum mechanics, but it has outlived its usefulness. Even though wave/particle duality complicates the effort, it seems to me that replacing the old icon with one better suited for the twenty-first century is a worthy challenge. Perhaps a public contest could be organized in connection with the celebration of the centenary of the birth of quantum mechanics in 1925.

4

The Wavefunction

The goal of physics is to explain the workings of the nonliving world. At first, philosophers described the properties of real objects: the wandering of planets across the night sky, the formation of ice, or the sound of a lyre. When attention turned to things that couldn't be seen or measured so easily, physicists invented mechanical models to take the place of real things. The Greek atomists substituted invisible particles moving through the void for continuous matter, Max Planck saw innumerable tiny oscillators in a ball of hot gas, and Niels Bohr imagined a microscopic solar system when he thought about the hydrogen atom.

Eventually, mechanical models failed too. They were duly abandoned, and replaced by much more abstract mathematical models. Compared to their predecessors, mathematical models are Spartan affairs. They consist of equations and formulas without the texture, the color, the visual detail—without the rich appeal—of their mechanical relatives. (Who can escape the endless fascination of dollhouses, model sailing ships, and model trains?) But what a mathematical model lacks in charm, it more than makes up for in generality and predictive power. Newton's law of universal gravitation was for centuries the reigning example of a purely mathematical description of a natural phenomenon.[1] It has withstood the futile efforts of generations of professional and amateur physicists to put flesh on its bare bones by "explaining" how the mechanical push of invisible particles or the swirl of some universal fluid "causes" gravity. And yet—what a boundless wealth of astronomical and terrestrial information is compressed into that little parcel of eight symbols ready for the unpacking by those who know how to read its message.

When the time came to develop a theory of the atom, the traditional categories turned out to be inadequate. The orbits and speeds of electrons in the outer shells of atoms were found to be inaccessible; atoms emitted light waves that showed up as particles; electrons acted like waves. Atomic physics upset the applecart.

Realizing that no mechanical model could convincingly imitate wave/particle duality, a handful of inventive physicists ushered in the quantum revolution by turning to a mathematical model instead. Their aim was to capture in mathematical language the strange facts that experiments in atomic physics revealed, without appeal to a picturesque description of the underlying reality. It was a bold move, and many of their colleagues found it hard to swallow.

But mathematical models of quantum phenomena bore spectacular fruit.

The big leap was to separate the object from its description. "Let's not look at the electron itself," the inventors of quantum mechanics cautioned, sometimes in so many words and more often implicitly. "Let's not even try to imagine a device that *acts* like an electron. Instead, let's search for a set of mathematical equations that predict how an electron behaves in the laboratory. The math doesn't have to look anything like a wave or a particle or even a platypus." And to their own delight, they succeeded.

The device that did the trick was a formula whose inventor Erwin Schrödinger (1887–1961) called the *wavefunction*. (Its spelling has evolved from the phrase *wave function* through the hyphenated *wave-function* to the more compact word *wavefunction*, which imitates its original German construction.) The wavefunction not only encodes the properties of a particular quantum system but also includes the essential details of the specific experiment that is performed on that system. So there's not just one wavefunction but a separate one for every distinct laboratory setup. In most cases a graphical representation of the wavefunction does not resemble a wave at all. Only its name continues to remind us of the one crucial property that quantum systems have in common: the possibility of superposition and of constructive or destructive interference—the ability of two waves to occupy the same spot and even to cancel each other out.

The mathematical form of the wavefunction is usually quite complicated—much more so than the equations $E = mc^2$ or $e = hf$. For that reason I won't display any examples of actual wavefunctions. But that doesn't mean we can't talk about them. You don't have to be able to read music to enjoy it.

An analogy even bolder than Bohr's image of the hydrogen atom as a miniature solar system inspired the construction of the wavefunction. For classical physicists one of the most perplexing problems of atomic physics was the discreteness of atomic energies. Unlike Earth satellites, which can orbit the globe at any distance and carry any arbitrary amount of energy, electrons confined in atoms are found only with definite, discrete energies. Where does that restriction come from?

The best example of the emergence, as if by magic, of discrete values out of a continuum is music. It has been known since time immemorial that musical instruments such as lyres, drums, and flutes produce individual fundamental tones, along with their overtones. When waves are confined to a restricted space—a fixed length of string, a circular drumhead, the hollow interior of a flute—they generate sound with a pure pitch, where you might have expected only noise. Pitch corresponds to the frequency of the sound wave that carries the note, and music is made by combining distinct frequencies. The question is this: In view of the fact that an atom doesn't resemble a flute except that it confines electrons, while a flute confines vibrating air, how can the well-known discreteness of *frequencies* in musical instruments help to explain the mysterious discreteness of *energies* in atoms?

The answer, of course, was suggested by the first tentative precursor of quantum theory, the fundamental connection between energy and frequency that is expressed in the celebrated Planck-Einstein relation $e = hf$.

The challenge for the inventors of quantum mechanics was to find a mathematical formula for a wave with discrete frequencies—inspired by the well-known formulas for the sound waves produced by musical instruments—which, via the relation $e = hf$, would match the energy levels of an atom. Such a formula would not describe the atom itself but would predict the observable staircase of its energy levels. Erwin Schrödinger succeeded in finding a general procedure for setting up a mathematical equation whose solution, in turn, is his celebrated wavefunction.

Quantum theory can be thought of as the science of constructing wavefunctions and extracting predictions of measurable outcomes from them. Over time, sophisticated techniques for doing that have evolved, first with the help of slide rules and later with computers. The systems that have been studied in this way have progressed from individual particles and atoms to bulk materials, to the interiors of stars, and even to the entire early universe. To date, quantum mechanics has passed every experimental test with flying colors.

The first system to be treated quantum mechanically was not an atom or even an electron but none other than the device that started it all—the harmonic oscillator. Its mathematical description involves only its mass and its unique and unvarying frequency. (The strength of the spring that forces the mass back to its resting place can be deduced from those two quantities, so it doesn't need to figure in the formalism explicitly.) As expected, Planck's constant h, the hallmark of quantum mechanics, plays a key role in the calculation. It sets the scale of things the way a meterstick, discreetly displayed off to the side of the picture, sets the scale for an archaeologist's photograph of a freshly opened trench.

As a theoretical guinea pig, the oscillator had the advantage of stark simplicity, but its shortcoming was the fact that in the twentieth century there were no actual mass-and-spring oscillators small enough to show the effects of quantum mechanics.[2] At best, the quantum-mechanical calculation served as a warm-up exercise for more difficult projects, such as a description of the hydrogen atom, which followed quickly and agreed with laboratory measurements. Nevertheless, even the mechanical oscillator illustrates some of the unusual departures of quantum mechanics from ordinary Newtonian mechanics.

Planck's desperate guess, that a harmonic oscillator's energies are multiples of $e = hf$, turned out to be almost right, but not quite. Surprisingly, the staircase of allowed energies does not start at ground level. Instead, the lowest energy is half a quantum, and the allowed energies are its odd multiples $e/2$, $3e/2$, $5e/2, \ldots$ Planck was lucky because the *differences* between energy levels, which determine how much energy a particular oscillator radiates or absorbs, are indeed multiples of e. That was all he really needed to assume. A quantum oscillator can no more radiate or absorb an energy of, say, $46.7\ hf$ than a grocer can accept a payment or give change of 46.7 cents in cash. It just can't be done! And if you try to drain the oscillator of all its energy to make it stop, you will fail. Like a hyperactive toddler, it never stops fidgeting. Remember, though, that because h is so tiny, an oscillator's residual tremor, after it has been deprived of all the energy it can possibly give up, is difficult to detect. Nevertheless, experimental evidence confirms this peculiar prediction of quantum mechanics.

Besides quantization of energy, the wavefunction implies superposition. According to classical physics, the position and speed of an object are always sharply defined. In contrast, the values of position and speed encoded in the wavefunction of an oscillator, or of any particle for that matter, can be spread out over a range—a superposition—of different values simultaneously. Notice that I did not claim that the position and speed of a particle can be spread out. The correct statement is: the position and speed *encoded in the wavefunction* can be spread out. That's an important distinction, and I'll say more about it in a moment.

The wavefunction is a little bit like a map—the best possible kind of map. It encodes all that can be said about a quantum system. I should mention here that the information contained in an ordinary map doesn't necessarily have to be displayed as a diagram on a sheet of paper or a globe. Road atlases, for example, often include a spreadsheet that lists the distances and driving times between cities. To simplify matters, imagine that the distances in the spreadsheet are not actual miles along real highways but are measured instead along straight lines "as the crow flies." Imagine an expanded version of this table for ten thousand towns in the United States. In principle it is easy to reconstruct the entire conventional map from the spreadsheet. Here's how: Place St. Louis in the middle of the page, place New York near the right-hand margin, and look up the distance between the two on the spreadsheet. Now you know the scale: how many miles correspond to an inch on your map. Then, from the spreadsheet, find the distances between both those cities and Miami and convert them to inches. Since the three sides of a triangle fully determine the tri-angle, you will now know where to position Miami. Continue for every other town to assemble the entire map. Astronomers record maps using a third method by listing the coordinates of millions of stars in fat catalogs without bothering to enter them on charts or globes. Maps, spreadsheets, and catalogs can be used to record the same sets of data. Though they look different, for many purposes they are equivalent. In the same way, the information contained in a wave-function can be displayed using a formula, a spread-sheet, a list of numbers, or a graphic image.

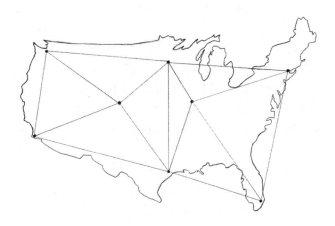

The first quantum mechanical description of an oscillator was, in fact, couched in terms of spreadsheets, which mathematicians call *matrices* (plural for *matrix*). Those matrices, in turn, promptly proved to be mathematically equivalent to wavefunctions. Since the latter are easier to imagine than the former, I'll stick for the most part to wavefunctions.

One of the most common traps that people—even some professional physicists—fall into when dealing with the puzzles of quantum mechanics is forgetting the difference between an object and its representation. The philosopher Alfred Korzybski memorably expressed that distinction in a different context when he coined the maxim "The map is not the territory." The phrase is a pithy reminder of the obvious fact that the description of an object is not the same thing as the object itself. Reality is not the same as a model of reality, any more than the word *house* is the same thing as a real brick-and-mortar house. Korzybski warned of the mischief that arises when the map is confused with the territory. Applied to quantum mechanics, his remark raises the suspicion that some of the strangeness of quantum mechanics might reside in the wavefunction rather than in nature. Could it be that it's the map that's weird and not the territory?

As children we learn to read maps by exploring the relationship between a street map and the asphalt and concrete it represents. What goes through our minds when we look up from the static little two-dimensional image and try to square it with the big, turbulent three-dimensional world around us and conversely, when we look down to sketch a simple schematic diagram of the complex real-world scene before us? That process of comparing the map with the territory is so difficult that some people never quite get the hang of it. The inclusion of motion, as on a car's GPS screen, confuses some people even more. A similar barrier to understanding has hampered quantum mechanics. In the quantum world, the Schrödinger wavefunction serves as a kind of evolving map constructed on a theoretician's laptop. If it's like a map, what exactly is it supposed to depict? How is it supposed to relate to the actual atomic landscape?

5

"The Most Beautiful
Experiment in Physics"

The wavefunction is a mathematical formula that encodes information about a quantum system. The quantum oscillator's wavefunction reveals that the little machine stores discrete amounts of energy—unlike a common tuning fork, which holds an arbitrary amount of energy depending on how hard it is struck. The wavefunction of the hydrogen atom similarly implies that energy is restricted to discrete steps or levels, but the level scheme is much more complicated than that of the oscillator.[1]

Besides predicting energy levels, the wavefunction predicts the outcomes of countless other experiments on a quantum system. The well-oiled mathematical machinery of quantum mechanics includes recipes for constructing the wavefunction for any conceivable experimental setup and instructions for calculating the results of measurements and observations. But instead of tackling such technical matters, let's try to come to grips with the meaning of the wavefunction by returning to the mystery that started all the fuss—the wave/particle duality of an electron. Let's see how the wavefunction manages to deal with that enigma.

Compare how physicists describe the flight of two very different projectiles—a rifle bullet and an electron.

First, the bullet. To make it simple, ignore gravity and air resistance. Once the bullet leaves the barrel, there are no more forces on it, and so, according to Newton's law of motion, it will continue in a straight line at constant speed until it hits its target, which we'll assume is made of wood.[2] There, it will suddenly encounter a braking force that, again according to the law of motion, slows it down until it stops. After stopping the bullet is squeezed from all directions, yet it experiences no net force and remains at rest—still in agreement with Newton's law.

The accuracy of the shot depends on the shooter and her equipment. The legendary sharpshooter Annie Oakley, it is said, could reliably hit a dime tossed into the air. Today, aided by elaborate and outrageously expensive equipment involving lasers, lenses, and laptops, even amateurs can beat her. Classical physics poses no limits to marksmanship. If the position and speed of the bullet at the moment of firing are determined within certain limits, its point of impact can be predicted within corresponding limits. In principle, though not in practice, that accuracy could be perfect.

With a sufficiently good rifle, a sharp enough eye, and a really steady hand, Annie could have hit any chosen spot on the dime.

And now the electron. It is shot from a device called an *electron gun*. Did you know that those weapons used to be more common in American homes than hunting rifles? Electron guns were essential components of old-fashioned TV sets, where they were hidden inside the tail end of the picture tube. You don't see them around much these days because flat-screen sets don't use them. Ignoring, as before, all intervening forces, consider the path of an electron from the gun to the screen, where it stops and produces a visible dot.

The quantum physicist, unable to track the electron directly, calculates its wavefunction instead. To do that he needs to know the geometric details of the electron gun as well as the speed at which the projectile leaves the muzzle. A graphic representation of the wavefunction, in contrast to those of harmonic oscillators and electrons in atoms, actually does resemble a wave emanating from the gun and traveling toward the screen. Like the wave a stone makes as it is dropped into water, the wavefunction spreads out as it advances toward the screen. Once the electron hits the screen, a miracle occurs. The wavefunction suddenly and inexplicably collapses to a point on the screen. Just before the hit, it is broadly spread out in space;

after impact the numerical value of the wavefunction is negligibly small everywhere except at the tiny dot that marks the arrival of the electron.

This phenomenon, called the *collapse of the wavefunction,* points the way toward the meaning of the wavefunction. Its flaw, which we'll take up in the next chapter, is its weirdness.

If the electron gun is fired over and over again, it paints a pattern composed of individual dots on the screen. The pattern offers the crucial clue for understanding the meaning of the wavefunction. The dots marking the arrival of electrons are made at random locations within the pattern. Random means without reason—unpredictable—lawless. That little word *random* describes a key difference between ordinary classical mechanics and quantum mechanics.

Of course, Annie Oakley would not have been surprised. After adjusting for atmospheric conditions, the peculiarities of her rifle, and her own pulse, she hit the dime every time—but her hits were distributed randomly over its surface. "There's no way to do better than that," she might have believed. The classical physicist, however, insists that the path of the bullet is predictable to any desired level of accuracy—provided the details of the entire system are known to the required accuracy. In classical physics only ignorance of the fine details or lack of control over them causes statistical randomness, which I call *Annie Oakley randomness.* In principle, though not in practice, randomness is absent from classical physics. Coin tosses, for example, are supposed to be truly random, but the tosses of mechanical coin flippers can be predicted. Annie Oakley randomness can be eliminated—not absolutely but as close to perfectly as you wish and can afford.

In sharp contrast, the randomness of electron gunfire is unavoidable. After the appropriate error bars on the dimensions of the gun and the speed of the electrons have been incorporated in the description of the experiment, the spreading of the wavefunction imposes an additional source of unavoidable randomness. In the early days of quantum mechanics, this *quantum randomness* proved hard for the physics community to accept. Einstein never made his peace with it—it was contrary to everything he had learned about physics during his long and spectacularly successful career. It "smelled" wrong to him, and since his keen scientific intuition had rarely let him down, he defiantly voiced his doubts about the emerging quantum theory to which he had contributed so much and which was racking up successes at such an astonishing rate. His ingeniously argued objections would keep physicists laboring to prove him wrong for years after his death. They succeeded—true randomness does exist—but a few of his most loyal defenders still hope that he may be vindicated in the end.

Quantum randomness (also known as *essential* or *intrinsic* randomness) violates a law that has been a cornerstone of physics since Aristotle—the law of cause and effect. Every effect is supposed to have a cause. Often, the cause is difficult to determine, but it is assumed to exist nevertheless. Thus, if Annie Oakley's bullet hit the L rather than the Y on a dime, we imagine that with sufficient effort we would be able to find the exact cause of the error. The electron, on the other hand, obeys quantum rules that absolutely deny this possibility. To a classical physicist like Einstein, rejecting the law of cause and effect seemed tantamount to pulling the rug out from under the enterprise of physics itself. We will discover how QBism places physics on a different, more resilient foundation that tolerates intrinsic randomness.

The pattern of dots made by an electron gun points the way toward understanding the meaning of the wavefunction. If the hits were entirely unpredictable, irregular constellations of dots might cover the entire screen. We would know nothing at all about the electron paths. But we do know something—in fact quite a lot. The wavefunction accurately describes the round, symmetrical bull's-eye within which the dots are concentrated and even the diminishing density of dots as you move outward from the center. So the electron gun presents us with an example of randomness mixed with partial knowledge.

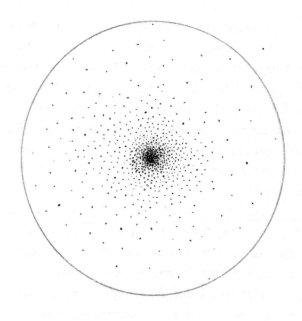

Such a mixture is the rule in science. Absolutely certain knowledge and complete ignorance are the exceptions. For example, error bars accompany every physical measurement. Even in everyday life, the extremes of both absolute certainty and perfect randomness are rare—just think of weather prediction and traffic patterns. In both cases we can predict a lot but not every detail. The mathematical tool for dealing with such situations is *probability,* a notion that is as central to quantum mechanics as Planck's constant h. The concept of probability, in turn, will turn out to be surprisingly tricky to come to terms with.

Looking at the screen in front of an electron gun suggests that the wavefunction does not describe electrons but probabilities. In particular, the wavefunction evaluated an instant before impact determines the probability that an electron arrives at a given spot on the screen.

The interpretation of the wavefunction in terms of probability is the real game changer that quantum mechanics imposes upon physics.[3]

We saw in Chapter 3 how the double-slit experiment with photons demonstrates the blend of randomness with lawfulness: within the pattern of stripes accurately described by the interference of waves from two separate sources, individual photons are recorded as dots randomly sprinkled over a photographic plate.

In his monumental textbook *The Feynman Lectures on Physics,* published in 1965, the year I started teaching physics, Richard Feynman began his discussion of quantum mechanics with a detailed, though hypothetical, description of the double-slit experiment performed with single electrons instead of photons. On the left is an electron gun, in the middle a minuscule double slit, and on the right a fluorescent screen that produces a dot when an electron hits it. In 2002 readers of the British journal *Physics World* voted this experiment "the most beautiful experiment in physics."

Even before Feynman's book was published, preliminary versions of it had been performed. But it wasn't until 2013 that technology had matured to the point that Feynman's thought experiment was actually put into practice almost exactly as he had described it nearly half a century earlier. Besides the difficulty of producing and detecting individual electrons, a daunting practical impediment had been the size of the slits. In the modern version, they are of nanometer dimensions (1 nm = 10^{-9} m = a billionth of a meter = a millionth of a millimeter), a feat of engineering that cannot be replicated at home with wire and electrical tape. A video showing the slow emergence of the striped pattern from randomly scattered dots is a mesmerizing experience of watching quantum mechanics at work.[4]

In addition to demonstrating wave/particle duality and quantum randomness, the experiment convincingly illustrates the spreading of the wavefunction. Each slit is about sixty nanometers wide. This number represents our ignorance of the electron's exact lateral position at the outset of its journey. The entire pattern of stripes on the detection screen, on the other hand, measures about three hundred micrometers from end to end. In order for the two portions of the wavefunction to overlap and interfere, they must each have grown in width by a factor of five thousand on their way from slit to screen. The wavefunction evidently spreads out considerably.

In contemplating this experiment, it is so easy to fall into error! Beams of light from a laser pointer directed at a double slit spread out, interfere, and produce striped patterns. Our minds inadvertently substitute streams of electrons for laser light and fail to be impressed. What we must not forget is that the electrons pass through the apparatus one by one. So feeble is their flow that if you removed the double slit and the screen in the 2013 experiment and just pointed the electron gun out the window, the electrons would follow each other into the air like ducklings waddling in a row, but they would be separated from each other by about two thousand kilometers. Each single electron is strictly on its own. The double slit splits only the wavefunction, not the electron, into two interfering portions. And yet each electron, far from the influence of its companions, somehow manages to avoid landing in the forbidden, unpopulated stripes on the screen as if guided by an unseen force.

The team who performed the modern version of the double-slit experiment calculated the electron wavefunction with great care, based on the measurements of their apparatus, error bars and all. This realistic computation is much more complicated and tedious than the idealized, simplified calculation found in Feynman's textbook. After measuring the positions of many thousands of electrons on the screen, the physicists compared the result—a striped interference pattern of otherwise random dots—with the quantum mechanical calculation. Their laconic final remark justified their herculean effort: "We see exactly what quantum mechanics predicts."

Feynman called the enigma of the double slit the "only mystery" of quantum mechanics. That's a bit of hyperbole because some quantum effects cannot be explained as "mere" wave interference, as we'll soon discover. Nevertheless, Feynman, who was not only a great physicist but also an inspiring teacher, managed to enshrine the electron double-slit experiment as the prototypical example of quantum mechanics in action.

6

Then a Miracle Occurs

If watching isolated electrons slowly painting striped patterns on a screen is an unsettling experience, thinking about the collapse of their wavefunction is even more bewildering. To see why, the comparison of a rifle with an electron gun is helpful once more. One moment a bullet flies smoothly at a steady speed; the next it enters the target and stops suddenly. Similarly, the electron wavefunction expands forward, subject to the rules of quantum mechanics, then changes character suddenly as a dot appears on the screen. Some similarities exist between the two scenarios, but the difference, though not immediately apparent, is striking.

Before, during, and after the rifle shot, the bullet never ceases to follow Newton's universal law of motion.

The wavefunction is less obedient. Before an electron hits the screen, its wavefunction evolves in time, spreading out as smoothly as a wave on a calm lake. The quantum mechanical law of motion fully predicts its development. Accordingly, the probability of finding the electron at a specific location spreads out over a rapidly expanding region in space. But when the electron stops on the screen, its description—its map—instantaneously and radically changes character. The wavefunction collapses as probability turns into (almost) certain knowledge of where the electron is located. The process of the collapse follows no rule or law. It just happens. Exactly why or how has been the subject of controversy since the birth of quantum mechanics ninety years ago.

In their search for a solution to the wave/particle puzzle, the inventors of quantum mechanics found themselves forced to compromise. By introducing the wavefunction, along with its probability interpretation, they succeeded in unifying wave-like behavior with particle-like behavior—but they had to pay a price. They had to give up a notion firmly embedded in the thinking of classical physicists from Newton to Einstein: the conviction that there is a unique law of motion for a material particle. It turned out that the electron's wavefunction does not follow one unchanging law of motion, the way a bullet does. Instead, the wavefunction obeys two fundamentally different laws:

1. As long as the electron is left to its own devices unobserved, its wavefunction unfolds smoothly, continuously, and predictably. It obeys fixed mathematical laws, just like a bullet in flight and a wave on a lake.

2. When the electron reveals its whereabouts by leaving a dot on a screen, the wavefunction "collapses" suddenly into a new, much more compact form concentrated on the point of impact.

The inimitable science cartoonist Sidney Harris has captured the situation with perfect understatement in his *Miracle* cartoon. I like to imagine that the two physicists are discussing quantum mechanics.

Wavefunction collapse is not only a "caving in" in space but a more general transformation of probability

"I THINK YOU SHOULD BE MORE EXPLICIT HERE IN STEP TWO."

into certainty. Not only position but also energy, speed, direction of motion, and many other attributes of a quantum particle, all of which have definite unique values in classical physics, can be spread out in the wavefunction over different possibilities, until a measurement is made, and a single value is unambiguously picked out. A wavefunction describes an electrical current, for example, flowing in opposite directions through a wire at the same time; a molecule with different geometric structures; and a radioactive nucleus that is simultaneously intact and decayed—until the question is asked and answered: Of all the *possible* things that could have happened, what *did* happen?

Miracles are not supposed to play a role in science. However, the world presents us with such an abundance of things we do not understand—such a boundless ocean of ignorance—is it really so surprising that the occasional miracle manages to sneak into scientific thinking after all? Only we don't call it a miracle. Newton's law of gravity is a perfect example.

Hold an apple in your hand. Let go of it. It falls to the ground. Why? Wouldn't it be more natural if the apple simply stayed put? If you were an astronaut floating in outer space, it would do just that—it would hover in front of you, right where you let it go. But down here on Earth, it falls.

Newton explained that the earth exerts a mysterious force called *gravity* that attracts the apple and pulls it inexorably down toward the ground. What are these invisible tentacles? Are they real or imaginary? What are they made of? How can we manipulate them, or even cut them off, to shed light on their nature?

Newton compounded the mystery by generalizing his law and claiming that all material objects, including your apple and the earth, exert this attractive force on each other. That's what keeps the moon in its orbit, the earth circling the sun, and you and me from floating off into outer space. *Universal gravitation,* it came to be called, and it's the prime example of an *action at a distance.*

But action at a distance is altogether unreasonable. Daily experience suggests that direct contact transmits forces. If you want to move a chair, you must touch it, directly with your hand or indirectly by means of a stick or a rope. Baseballs reverse direction when they touch the bat, not before or after. Molecules jiggling their nearest neighbors and passing on their motion in a kind of chain reaction transmit sound and heat. Photons carry light and radio waves from their sources to their receivers. On a microscopic scale, even so-called contact forces such as the push of a hand or a baseball bat are ultimately mediated by electric and magnetic fields that transmit disturbances from point to neighboring point. Only action at a distance dispenses with the universal need for closeness between interacting bodies. It's a miracle masquerading as a "law of nature."

Think of your own effect on the universe. According to Newton, when you move your body by taking a step forward, every atom in the universe, every person on Earth, every planet and every star, no matter how distant, instantly experiences a change in the gravitational force on it. It's as though the distant matter somehow reacts to what your body is doing, instantly and without a messenger to carry the message.

Newton, of course, understood the implausibility of his own law. Years after universal gravitation had been enthroned as a great law of nature, in a letter to a correspondent who wondered about the action-at-a-distance view of gravity, Newton wrote:

> It is inconceivable that inanimate Matter should, without the Mediation of something else, which is not material, operate upon, and affect other matter without mutual Contact.... That Gravity should be innate, inherent and essential to Matter, so that one body may act upon another at a distance [through] a Vacuum, without the Mediation of any thing else, by and through which their Action and Force may be conveyed from one to another, is to me so great an *Absurdity* [emphasis added] that I believe no Man who has in philosophical Matters a competent Faculty of

thinking can ever fall into it. Gravity must be
caused by an Agent acting constantly according
to certain laws; but whether this Agent [is] material or immaterial, I have left to the Consider-
ation of my readers.[1]

An absurdity, Newton calls his own greatest contribution to physics! He had created it around 1666,
when he was twenty-four years old. That period of his
life is known as his annus mirabilis, his miracle year,
not because action at a distance is indeed a miracle but
because in that year Newton, in a miraculous fit of cre-
ativity, also invented calculus and dissected sunlight
into the colors of the rainbow.

A quarter of a century after proposing it, Newton,
far from repudiating action at a distance on the
grounds of its unreasonableness, defends it for its usefulness but admits that it is ultimately incomprehensible. He has figured out by what law gravity operates
but not what it means. As a devout believer, he privately
ascribes the action of gravity to God but wisely leaves
it up to his readers to draw their own conclusions. He
describes the miracle in succinct mathematical terms,
but he cannot explain it.

And yet the law of gravity reigned unchallenged
for a full quarter of a millennium, from 1666 to 1916,
when Albert Einstein discovered its true nature. To
be sure, the intervening years brought countless

attempts to explain gravity in terms of complicated mechanical models, but none stood up to experimental or mathematical scrutiny. For 250 years physicists used Newton's law of universal gravitation to explain the world and to make wonderful predictions, ranging from ocean tides and the flattened shape of Earth to the timing of solar eclipses and the reappearance of comets. So successful was the absurd law that its form was copied in the mathematical treatment of many other phenomena unrelated to gravity, such as magnetic and electrical forces.

Einstein objected to action at a distance because it violates not only common sense but more importantly to him, the special theory of relativity. In 1905, his own annus mirabilis, he had proposed that no object, no signal, and no information can travel faster than the speed of light. Action at a distance, on the other hand, travels through the void with infinite speed—an impossibility according to relativity. So Einstein developed his own theory of gravity and called it *general relativity*. General relativity replaced action at a distance by showing how space itself acts as the medium for transmitting a gravitational force from point to nearby point. Such a process—the opposite of action at a distance—is called *local action* because an influence located at a point in space affects only points in its own immediate, local vicinity, not points at a distance. On this view if you take step forward, the space around your body is subtly warped a little bit, and that disturbance travels out, point by point at light speed, to the far reaches of the world, the solar system, the Milky Way, and the universe. After 250 years the miracle of gravity had finally been replaced by something much more complicated but at the same time much more explicit.

Newton's venerable old theory was reduced to the status of an approximation; a very useful approximation to be sure but a concept without fundamental significance. Physicists use it in the same way they approximate solids, liquids, and gases as continuous materials even though they know that matter is really composed of atoms.

The collapse of the quantum wavefunction, which covers arbitrarily large distances in an instant, is also an action at a distance, and it is just as incomprehensible as Newtonian gravity. But by proving its worth as convincingly as Newton's law did, the collapse of the wavefunction has also made its way into scientific orthodoxy. The great majority of physicists accept quantum mechanics as proven fact—superposition, probabilities, wavefunction collapse, and all. "That's how nature behaves!" they say to themselves and get on with their calculations and observations. Only a small, though growing, number of them take seriously the philosophical conundrums implied by the standard formalism and try to resolve them. One of the principal goals of those intrepid souls is to become more explicit in step two of the quantum recipe, the wave function collapse, which takes them in an inexplicable leap from probability to certainty.

7

Quantum Uncertainty

Werner Heisenberg's *uncertainty principle* has become a meme of popular culture almost as famous as Einstein's $E = mc^2$ and Schrödinger's cat. From the bumper sticker "Heisenberg may have slept here" to the "Heisenberg" alias of Walter White, the modern-day Dr. Jekyll and Mr. Hyde of the TV drama *Breaking Bad,* Heisenberg's name evokes the notion that quantum physics has eroded the certainties of yesteryear. But the interpretation of his principle as a claim that "everything is uncertain" is a superficial mistake. More consequential than this common misinterpretation was an error that Heisenberg himself committed. The principle that bears his name is a mathematical theorem derived from the wavefunction, and it is impeccable. It states that the position and velocity of a particle cannot be completely specified at the same time: the more precisely the position is determined, the more uncertain the velocity becomes—and vice versa. The precision of other pairs of variables, such as energy and duration, is subject to similar trade-offs. But Heisenberg's explanation of the deeper meaning of his mathematical theorem was flawed.

Heisenberg's theorem is a blunt instrument. Though it rarely figures in detailed calculations, it serves as a useful rule of thumb. It allows quick and rough estimates of properties of atomic systems before the full theory yields more reliable answers. For example, the uncertainty principle helps to make sense of the lowest rung on the energy staircase of a quantum oscillator. Assume, incorrectly, that the lowest energy is precisely zero, so you know both the speed and the displacement of the little mass: both are exactly zero. The mass is at rest, and the spring is relaxed. Since your assumption violates the uncertainty principle, it must be wrong. If the oscillator is to obey the rules of quantum mechanics, its mass must jiggle a little bit, with both position and speed varying and therefore uncertain within limits. A hand-waving argument based on the uncertainty principle even shows, correctly, that the lowest energy of the quantum harmonic oscillator is not zero but $e = hf/2$. Unfortunately, you can't trust this estimate until it is verified, with much greater effort, by a meticulous computation of the actual wavefunction.

Since the uncertainty principle runs so dramatically counter to the very foundations of classical mechanics, which endows every bullet and every golf ball with a precise position, speed, and direction of motion, Heisenberg undertook to explain the physics behind his theorem. This exercise was actually somewhat unusual for him—he normally preferred abstract, mathematical, Platonic considerations to realistic, intuitive, Aristotelian arguments. Nevertheless, he proceeded to illustrate his principle in ordinary, practical language that seemed convincing to generations of physicists, including myself. But in the end, his reasoning turned out to be misleading, even though the principle itself is correct.

Heisenberg sought the origin of quantum uncertainty in the effect of measurements on the object being measured. He devised a clever hypothetical experiment, *Heisenberg's microscope,* to illustrate the idea. "Consider an electron in flight," he suggested. In order to figure out exactly where it is, you have to catch it, or touch it, or bounce light, or at the very least one photon off it to gain information about its position. That photon, in turn, will knock the electron around a bit—changing its speed or direction or both. So while a deflected photon will help to locate the electron at some specific time, the observation changes its velocity. By going through the details of this imaginary experiment with great care, Heisenberg—after some initial missteps—was able to construct a plausible physical illustration of his uncertainty principle.

What he was invoking should be called an *observer effect,* a phenomenon that is both real and easy to understand. You don't need quantum mechanics to find examples of the effect of an observation on the observed object. Chemists know very well that inserting a room-temperature mercury thermometer into a thimble full of hot water will lower the temperature of the water. Lawyers know that the manner of their questions will influence the answers. Anthropologists take care to minimize the effects of their research on the culture they try to describe. And in the worst case, an observation can even destroy the object—an autopsy may reveal the cause of death, but it wrecks the body.

In the nine decades since Heisenberg announced his principle, the realization has slowly grown among physicists that it depends neither on the disruptive effect of physical measurements nor on the precision of measuring instruments. In fact, it is much deeper and follows from the wave nature of matter, of which the word *wavefunction* is a constant reminder. Even classical waves display a built-in reciprocal relationship between duration and frequency. Imagine a disturbance of the ocean surface made up of a ripple of waves. If it consists of several cycles, each with a crest and a trough, you can time them and determine their frequency. The entire ripple is extended in both space and time—its duration is long. On the other hand, if the ripple is composed of a single swell, its length and duration may be much shorter, but you can't define the frequency because you need at least a full cycle to do that. At best you can regard that solitary wave crest as a jumbled superposition of many waves with different frequencies, all of which happen to crest together at the peak of the swell. The trade-off for classical waves implies that the longer the duration of a wave, the smaller its spread of frequencies and vice versa.

This reciprocal relationship holds not only for water waves but for sound waves too, and its effect can be heard in an orchestral concert. The oboe's prolonged tuning note A, which starts the evening, has a single well-defined pitch or frequency, but a cymbal clash, which lasts only a fraction of a second, has no discernible pitch at all. In fact, the printed score for percussion instruments uses a special notation with no reference to pitch—because the pitch of a clash is undefinable. But its timing is unmistakable!

The Planck-Einstein equation $e = hf$ turns the classic trade-off between duration and frequency into the Heisenberg uncertainty relation between the lifetime and the energy of a quantum system such as an unstable particle. Here again, as in the derivation of the wavefunction itself, Planck's constant furnishes the link between classical and quantum physics.

The most drastic illustration of an uncertainty principle is the double-slit experiment. It displays the uncertainty between the wavelength and *which-path* information; that is, the answer to the question, "Through which of the two slits did the particle actually pass?" The wavelength can be easily deduced from the dimensions of the apparatus and the interference pattern.[1] Knowledge of which path is more difficult to obtain—except by brute force. If you cover up one of the slits, you *know* that the beam traveled through the other one. But when you do that, the interference pattern, and with it the evidence for wavelength, disappears. (Of course it does. It is created, after all, by the interference of two waves.) In this example the measurement of which- slit information is disruptive in the extreme: it eliminates one path altogether. So the uncertainty is also extreme—either the wavelength or which path can be determined with confidence but not both at the same time.

Even as a deeper understanding of Heisenberg's principle was developing, new technology was suggesting new ways of manipulating individual elementary particles and converting yesteryear's imaginary experiments into real observations in the lab—as it did for Feynman's beautiful experiment. Modern refinements made it possible to analyze the uncertainty of the double slit not just in the obvious old all-or-nothing version but for approximate knowledge of the wavelength and probable knowledge of the path. And that wasn't all. By the beginning of this century, new incarnations of the venerable experiment began to demonstrate Heisenberg's error explicitly: quantum uncertainty is not an observer effect.

The ingenious innovation is to separate the "observing mechanisms" for which-path observation by a safe distance so they can't possibly interfere directly with the particle—a photon in this case.[2] Immediately upon emerging from the double slit, each photon is sent into a special crystal, where it spontaneously generates two new photons with identical (or complementary) properties. These two are sent in different directions on different errands—one, called the *signal,* to contribute to the slowly emerging interference pattern (or its absence) in the usual way, the other to serve as witness. Each signal photon is linked to a unique witness photon.

The witness arrives at its destination after the original photon has passed the double slit, an arrangement that explains the name *delayed choice experiment.* The witness is interrogated by means of standard optical wizardry to reveal either which slit the original photon came from or whether it emerged from the two slits without divulging which path it took.

With this arrangement the signal detector records many thousands of photons as it scans over a broad area. Each detected signal photon corresponds to a dot on the screen of the old-fashioned double-slit experiment, only now each signal photon has a witness. Next, the experimenter faces a choice. First, from all the collected data, she selects only those signal photons for which, according to the witness, which-slit information is absent. Plotting the positions of the signal detector—the dots on the screen—she finds the expected striped interference pattern. In fact, she has reproduced Thomas Young's experiment of 1803. Second, alternatively, if she selects only the signal photons whose which-slit question has been answered and records their positions, no stripes appear. But both slits have remained wide open for both parts of the experiment.

The message of the outcome is clear. The witness detector clicks so far away in space and time that it can have no direct physical influence on what happens at the slits. The disappearance of the interference pattern is not a mechanical response to the observation of the path taken by the signal, as it is when a slit is blocked. In short, the uncertainty principle is not an observer effect.

The progress from Heisenberg's microscope to the interpretation of the uncertainty principle as a very basic, general property of wavefunctions is reminiscent of other developments in the history of quantum mechanics. Planck's mechanical model of glowing matter led to wave/particle duality and its resolution by the wavefunction. A purely mathematical wavefunction and its interpretation in terms of probability replaced Bohr's mechanical model of hydrogen. In each case a mechanical, visualizable description turned out to be inadequate, and a mathematical, abstract explanation replaced it.

Abstraction is a sign of maturity. Children begin to learn about money by handling coins, but later their understanding broadens to include abstract concepts such as cost, price, and credit. In society at large, the notion of justice evolved from the primitive, personal "an-eye-for-an-eye" principle to sophisticated systems of abstract laws. In physics, maturity implies pulling away from tangible mechanical models toward mathematical abstractions (Latin *abstrahere,* to pull away from). Things are concrete—thoughts are abstract. But abstraction should not be confused with complexity. A concept may be abstract, but it doesn't have to be complicated.

8

The Simplest Wavefunction

Start simple!" is good advice in most human endeavors, even in science. Niels Bohr started with hydrogen before venturing on to more complicated atoms; quantum mechanics cut its teeth on the simple harmonic oscillator. So let's consider the simplest possible wavefunction—not in terms of mathematical equations but in the form of a visual symbol. The exercise will illustrate four fundamental properties of wavefunctions—superposition, probability, discreteness, and collapse. As an added bonus, it will also turn out to be useful later on when we explore the implications of QBism.

Since atoms, even the simplest ones, are marvelously intricate structures, we'll look instead at a truly elementary, uncuttable particle. We have encountered two of these, the photon and the electron. Photons elude description in simple words: In a vacuum they are always flashing around at the speed of light, refusing to be slowed down or stopped for close inspection. When they are detected in some way, they give up their energy and vanish. To describe their ghostly properties, physicists must reach beyond wavefunctions and the language of ordinary quantum mechanics. Electrons, on the other hand, can be slowed down, stopped, stored, examined, and manipulated almost as easily as marbles, so they promise to be more accessible to everyday intuition. What's more, they're not only essential ingredients of matter, including our bodies, but as carriers of energy (in power lines) and information (in computers) they serve to fuel and organize our lives. The electron, that mighty midget, is an appropriate vehicle for focusing our ideas about the invisible microworld.

The description of an electron includes its position, velocity, mass or weight, and electric charge. In addition, an electron has two other related properties. The first is rotation about its own axis, called *spin,* and the second is magnetism.[1] An electron behaves like a tiny bar magnet, or a miniature compass needle, with an unvarying, well-measured magnetic strength. Quantum mechanics correctly predicts that strength with the mind-boggling precision of about one part in a billion. (That's roughly the ratio of the width of your thumb to the distance from New York to Hawaii.)

The list of attributes of an electron also applies to a spherical electrically charged pellet made of, say, plastic. When such a little ball rotates about its own axis, it also behaves like a bar magnet. So we are tempted to think of the electron as a miniature version of a terrestrial globe. It is simpler than the earth because it is perfectly spherical, and furthermore its two axes—the rotational or spin axis and the magnetic axis—coincide. (In contrast to our planet, the magnetic poles of a spinning pellet are also its geographical poles.) But quantum mechanics is more than just the classical mechanics of small stuff. Careful examination of the electron will lead us right out of this world into an alien dimension.

Did you notice that the aforementioned list of properties of the electron includes mass but not size? So how big is an electron? Or rather, how tiny? The surprising answer is that no measuring instrument, no matter how accurate or complex or expensive, has ever detected an electron size. More to the point (pardon the pun), when theoreticians introduce a minuscule hypothetical electron radius into their equations, many successful predictions, including that of the magnetic strength, are thrown out of whack. The best assumption, which consistently leads to phenomenally accurate predictions of experimental outcomes, is that the electron's radius is zero. As far as we know, the electron is a point particle. Of course, one day we may find out that the electron does have a substructure and a radius after all, and then current theories will need to be refined—but to date that's just speculation. So let's go with what's known and try to imagine a particle with no size at all!

The trouble is that a point particle cannot rotate. A point can whirl around in a circle, but it is impossible to make sense of the notion of a point spinning about its own axis. Spinning implies that different parts of the object move in opposite directions, so a point, having no parts, cannot spin. Baseballs and ice skaters can do it, but a point is too insubstantial to spin. The mechanical model of the electron as a charged, spinning ball is therefore untenable. The word *spin* itself is a misleading conceptual fossil of the same vintage as the Bohr model of hydrogen. Unfortunately, we're stuck with the paradoxical conclusion that the electron possesses spin and magnetism but no size.

Trying to apply concepts from our macroscopic world to the microscopic realm of the quantum led us into trouble when we discovered wave/particle duality, and it has done so again. In order to regain some peace of mind, we have to reach deeper into our imagination. Perhaps we can learn from Alice in Wonderland's encounter with the Cheshire Cat. As its body grows fainter and finally disappears, the cat leaves nothing behind but its grin, prompting Alice to remark that she has often seen a cat without a grin but never a grin without a cat. From afar an electron looks like a rotating pellet that grows smaller and smaller until it disappears, leaving nothing behind but its spin.

Spin gets more puzzling still. An electron's spin, unlike that of a pellet, cannot be slowed down or speeded up. It has a fixed magnitude determined by the value of Planck's ubiquitous constant h. In order to find out which way the electron's built-in compass needle (and hence its spin axis) is pointing, you can bring it near the north pole of an ordinary refrigerator magnet. Left to its own devices, the electron will duly line up so its magnetic south pole points toward the magnet, and its north pole points away. You can turn the electron around, so it points the wrong way, but you have to expend a bit of energy to do that—like pushing a compass needle around with your finger.

Unlike an ordinary bar magnet, whose strength and direction can be changed arbitrarily, an electron's magnetism is fixed in magnitude and restricted in direction. In particular, when an electron's spin (and hence its magnetism) is measured, only two possible values can turn up. Every device used to measure spin contains a fixed external magnetic field to provide an arbitrarily chosen reference direction. Strangely, an electron's spin is always found either lined up with or against the reference direction. It never lines up perpendicular to the reference or at forty-five degrees to it even while it's being turned around. The electron's magnetism is often depicted by a little arrow, which also represents its spin. When the spin direction of an electron in a vertical magnetic field is measured, it points either up ↑ or down ↓, never at an angle to the vertical. Similarly, if the reference field is lined up horizontally along the x axis, the electron will only be found to point right → or left ←. Instead of a pellet's infinite range of spin directions, an electron has exactly two. Such restrictions on the range of orientations resemble the restriction on the energies of harmonic oscillators and of atoms, which in turn are reminiscent of the restrictions on the pitch of the sound of a flute.

Along with other variables that describe the atomic world, spin too is subject to an uncertainty principle or information trade-off. If you prepare an electron with its spin pointing up ↑, for example, and later measure the spin along the horizontal x axis, the result will be left ← or right → at random. Conversely, if you know that the electron is spinning right →, its orientation measured subsequently in the vertical direction will be randomly split between up ↑ and down ↓.

We have landed smack in the middle of the wonderland of quantum mechanics. Spin, with its peculiar rules, is a quantum phenomenon that does not play a role in the double-slit experiment, which Feynman called "the only mystery of quantum mechanics," but it is an enigma nonetheless.

The wavefunction of an electron involved in any experiment has two portions. The "external" part deals with motion through space—inside an atom, or from an electron gun to a screen, or through a double slit—the part we've dealt with up to now. In addition, there is an "intrinsic" part that deals only with spin.

Often, these two components of the total wavefunction are intertwined in the calculation, but for our purposes we separate them from each other, ignore the external wavefunction, and contemplate only the part that describes spin. With that we have arrived at our destination: the simplest possible wavefunction.

Unlike an ordinary wavefunction, which may be spread all over three-dimensional space and has an infinite range of values corresponding to places where the electron might be found, the spin wavefunction is not located in real space. The invention of the spin wavefunction, a purely abstract, purely quantum mechanical construct with no analog in our everyday world, represented one of the most revolutionary events in the early history of quantum mechanics. It implied that every electron has two hidden states; a kind of bipolar personality that only reveals itself when its magnetic field, or its rotational motion, is observed. Otherwise, the twofold character of the electron remains concealed in an alien dimension unrelated to the space we live in.

Electron spin is a keyhole through which we glimpse the quantum world, a world we fail to recognize not because its features are too small but because some of them are accessible only to our imagination, not to our immediate senses. Among the countless Einstein quotes that have become part of popular culture, one of the more reassuring is this: "The Lord God is subtle, but malicious he is not."[2] Leaving God out of it, the remark suggests that the secrets of nature are deeply hidden and difficult to tease out but ultimately accessible to reason and imagination. When nature presents us with an apparent paradox, she often obligingly whispers clues for its resolution into our ears. Electron spin is such a clue: it allows us to peer into the secret world of the quantum.

Since the word *spin* and its familiar associations with baseballs and ice skaters is a quantum mechanical misnomer anyway, the two observable states of the spin wavefunction don't have to be labeled clockwise and anticlockwise. In fact, they can be called up/down, as in the original mechanical model of the electron, or right/left, +/−, yes/no, heads/tails, on/off, or black/white, but in order to make contact with computer code they are conventionally designated by 0 and 1. These two integers are merely convenient labels, like page numbers.

The spin wavefunction is immensely useful beyond the context of electronic spin and magnetism because it describes any quantum mechanical system that has only two possible real configurations. It can refer to a molecule that switches back and forth between two different structural arrangements, to an electrical current that flows clockwise or counterclockwise in a wire loop, to an electron that can occupy one of two specific energy levels in an atom, to a light beam that is polarized horizontally or vertically, or to a radioactive nucleus that is intact or decayed. Exactly the same simple wavefunction describes these and countless other systems. On account of its simplicity, this mathematical object is beginning to replace Feynman's double-slit wavefunction as the starting point for college courses on quantum mechanics.

In the language of spreadsheets, electron spin is described by a 2×2 matrix—the smallest possible square matrix. (A 1×1 matrix doesn't really deserve that name. It's just a number and unable to display quantum superposition.)

Spin-like systems are so ubiquitous that they have earned a name of their own. Any quantum system with just two possible states is called a qubit, pronounced "cubit." Qubit is a contraction of *quantum bit* while the word *bit* itself is a contraction of *binary digit*. A classical bit is simply a quantity with a value of 0 or 1, an abstract symbol of a toggle switch labeled *off* and *on*. A qubit, in contrast, is a real physical quantum mechanical object or system. It's a thing, not a symbol.

Unfortunately, the word qubit has nothing to do with the word *QBism,* the topic of this book. Not only are their homophones cubit and Cubism, respectively a biblical measure of length and an early twentieth-century art historical period, utterly irrelevant to quantum mechanics, but qubits and QBism are also unrelated. They agree on the letter q for *quantum*, but the lowercase b means *binary*, whereas the capital B stands for Thomas Bayes, an eighteenth-century clergyman. Sometimes the topsy-turvy world of scientific nomenclature produces strange bedfellows!

A qubit is described using a mathematical device called the *qubit wavefunction*. In order to distinguish the qubit from its wavefunction—the territory from its map—in this book I will use the designation *qubit* as shorthand for the qubit wavefunction. The font intentionally underlines this distinction because in the professional literature Korzybski's warning is often ignored.

A point on a sphere can symbolically represent the *qubit* for a particular system involved in an experiment. Every point on the surface corresponds to a probability. At the poles the measurement outcomes,

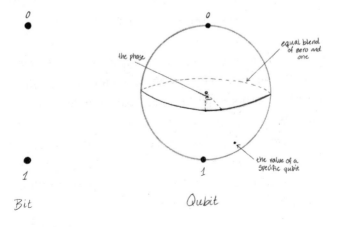

whatever they happen to be, are labeled 0 and 1. Between these extremes lie blends, or superpositions, of the two values. For example, an event whose *qubit* is located on the equator has a 50 percent probability of coming out 0, like heads in a coin toss. At latitudes in the northern hemisphere, the event is more likely to come out 0 than 1 and vice versa at southern latitudes. In contrast to the latitude, the longitude of the point on the sphere has no classical counterpart. It is a purely quantum mechanical variable and represents a phase, measured by an angle in an imaginary, inaccessible space. Two neighboring *qubits* on the sphere tend to interfere constructively (crest meets crest and trough meets trough), while those represented by points on opposite sides of the globe interfere more destructively (crest meets trough). The phase is the last echo of a classical wave, with its characteristic property of superposition, which inspired quantum mechanics in the first place and lent its name to the wavefunction.

Thus, the little *qubit* sphere is a visual reminder of the phenomenon of superposition and its interpretation in terms of probability. Except at the poles, a point on the sphere does not help to predict the outcome of an isolated measurement. Repeated trials of identically prepared experiments yield 0s and 1s in random sequence. The latitude of the point predicts how often each result appears in the sequence.

The exceptional points, the poles, which are not superpositions and don't have a phase, reflect the discreteness of quantum mechanics. Just as the energy levels of quantum harmonic oscillators and of real atoms are discrete and countable, rather than continuous, many other measurements, including the sense of an electron's spin, are restricted to a countable number of values—two for a *qubit*. The poles anchor the image in the real world. Taken together, they are represented by a bit.

Perhaps the most compelling visual message of the *qubit* comes from what it is *not*. It is not a picture of an electron nor, in contrast to Bohr's icon, of anything at all in our world. Its three dimensions are products of the imagination. A point on the surface of the little ball represents the probability of the outcome of an experiment, but after the experiment is actually performed the system is found to be in state 0 or state 1. A point on the sphere, in other words, jumps to a pole. This leap is the notorious collapse of the wavefunction.

A point on the sphere may be fixed in time, or it may wander on a prescribed path. Consider, for example, a radioactive nucleus produced at a specified time. Let the value of the *qubit* represent the answer to the question: Has the nucleus decayed, either by splitting or by emitting some kind of radiation? The answer no is represented by 0; the answer yes by 1. Initially, the point on the *qubit* ball is at the upper

pole labeled 0. As time progresses the probability that
the nucleus has disintegrated increases, so the point
slides down toward the lower pole. It will, however,
never get there as long as the nucleus remains unobserved. If you actually check the condition of the
nucleus, you will find that it is either intact or decayed. At that moment the *qubit* collapses onto one of
its poles. The journey of the point along the surface is
entirely predictable and mathematically described
by quantum mechanics but the instantaneous jump
down to the south pole or back up to the north pole is
not. After a measurement the *qubit* assumes a bit
value of 0 or 1, but before the measurement the *qubit*
has no bit value to be discovered.

The image of the *qubit* ball does not *explain* superposition, probability, discreteness, or wavefunction collapse nor does it reveal the mathematical formulas it symbolizes, but it serves as a compact visual
reminder of the principal ingredients of quantum
mechanics. It's a picture of the simplest possible
wavefunction, even though it doesn't look the least
bit like a wave.

II

Probability

9

Troubles with Probability

The rules of quantum mechanics are crystal clear in their instructions for constructing wavefunctions. Sometimes the recipe poses difficult mathematical and computational problems, but there is rarely any doubt about *what* to do—it's only the *how* that keeps physicists scratching their heads. At the end of their labor, they have in hand a wavefunction and are ready to take it down to the laboratory.

The link between theory and experiment turned out to be *probability:* either the wavefunction predicts probabilities and the laboratory furnishes the data to test them, or conversely, experimentally determined probabilities guide the calculation of the wavefunction, which then encodes information about other possible experiments and allows predictions to be made for those. At first glance the concept of probability appears to be so elementary that it is intuitively obvious. What's the probability of heads in a coin toss? One-half, or 50 percent, as every football captain knows. What's larger, the probability of rolling 6s or 7s with a pair of dice? Let's count the ways. Altogether there are $6 \times 6 = 36$ possible throws, but among them just a handful produce 6s or 7s: (1, 5), (5, 1), (2, 4), (4, 2), and (3, 3) versus (1, 6), (6, 1), (2, 5), (5, 2), (3, 4), and (4, 3). The two probabilities are $5/36 \approx 13.9$ percent and $6/36 \approx 16.7$ percent, respectively, so 7s are about 3 percent more likely than 6s, as a really attentive craps player knows from experience.

The probability for the occurrence of an event is simply the number of favorable outcomes (for example, six pips showing) divided by the number of possible outcomes (for example, thirty-six). Even if the number of events is not countable, this formula usually works. What's the probability that a blindfolded toddler pins the tail somewhere on the body of a donkey on a poster, assuming that the pinpricks are randomly distributed over the entire poster? Just divide the area of the donkey by the area of the poster. The result is a real number between 0 and 1—a valid probability, expressible as a fraction or a percentage.

Probabilities so computed are abstract, theoretical numbers. How they add up and combine in complicated scenarios is the subject of the branch of pure mathematics called *probability theory*. The probabilities the theory deals with are no more real than the infinitely thin lines, dimensionless points, and perfect circles of Euclidean geometry. Whether the abstractions of probability theory and Euclidean geometry have real-world applications is not a matter of logic but of experiment and observation—a question of science. We may feel that tossed coins and rolled dice are so simple that our intuitions about them need no confirmation, but like many things in life, the truth is more subtle. Best be prepared for surprises!

Consider the baffling paradox known as the *cube factory,* which was based on similar puzzles of older vintage and posed by the philosopher Bas van Fraassen in 1989. (The example seems singularly appropriate in the context of QBism!) Imagine a pottery factory that spits out a huge pile of small ceramic cubes with edges randomly distributed in length from 0 to 1 cm. You pick out one of these cubes at random and examine it. What is the probability that the edge of your cube measures between 0 and 0.5 cm? A tempting answer is "one-half" because the range of favorable outcomes is half the total available range. But wait! You notice that the area of each side of a cube varies from 0 to 1 cm^2. What is the probability that the one in your hand has a side measuring between 0 and 0.5 cm × 0.5 cm, or 0.25 cm^2? Since 0.25 is a quarter of the total range, the probability that your cube falls into that interval is "one-quarter." It gets worse. If you measure volumes instead of lengths or areas, they range from 0 to 1 cm^3, and the question becomes: What's the probability that yours measures between 0 and 0.5 cm × 0.5 cm × 0.5 cm = 0. 125 cm^3? The answer is "one-eighth." Three different answers to a simple question scream paradox. Which one is correct?

Mathematically, the problem has no solution. In a real case, one of the answers might be picked out by taking into account the actual manufacturing process. Somewhere inside the machinery there must be some kind of a randomizing procedure. Is it a caliper that is randomly varied to measure between 0 and 1 cm? If so, the first answer is correct. Or is it a scale that randomly weighs out blobs of clay corresponding to volumes between 0 and 1 cm^3, which are then formed into perfect cubes? In that case the third answer is correct. Or perhaps the randomization occurred in an altogether different way, yielding yet a fourth possible answer to van Fraassen's question.

The cube factory is a potent reminder that probability is a sharp mathematical tool that must be wielded with care when used in a real application.

Not only logic and mathematics but nature itself can produce surprises. Consider two balls, painted white and black, respectively, stored randomly in two different urns. There are only four ways to distribute them: (WB, 0), (W, B), (B, W), and (0, WB). The probability of finding both balls in the same urn is evidently 2 out of 4, or 1/2. This way of determining probabilities was for centuries the standard way of counting and looks perfectly obvious. It works for dividing votes among two candidates and figuring the odds for poker—but in the quantum world it turns out to be wrong.

Photons are not like balls in an urn. Their behavior displays another peculiarity of quantum mechanics that has no parallel in the everyday world: photons with the same frequency (the same color) are absolutely indistinguishable from each other. New pennies resemble each other too, but on a microscopic scale, their rugged surfaces are easily distinguishable. Even if the coins were identical as far as our technologically aided senses can determine, their journeys through space and time could be followed and used to tell them apart no matter where they might have wandered. Pennies can be distinguished by their history as well as their appearance. "This is penny A and that one is penny B" is always a valid, verifiable statement. Photons, however, cannot be labeled like that. Once they come close to each other, they express their wavelike character, overlap into superpositions, and lose their identities. Unlike pennies, they are fundamentally indistinguishable.

Distributed among two different polarizations (which stand in for urns), the only possibilities for assigning two otherwise identical photons, represented by asterisks, are (**, 0) (*, *), and (0, **). Now the probability of finding both photons in the same state of polarization has risen from 1/2 to 2/3. That increase may not seem like much, but when repeated a trillionfold in an actual application, it fundamentally changes the statistics of photons. The Indian physicist Satyendra Nath Bose first worked out the consequences of this unconventional way of counting and, by focusing on photons rather than hypothetical oscillators, succeeded in rederiving Planck's radiation law. Einstein, who had invented photons, was surprised and deeply impressed by this calculation. He made sure that other physicists heard of it and then generalized Bose's version of the resulting statistics to apply to massive particles as well as photons. Eighty years later the 2001 Nobel Prize honored the experimental observation of Bose-Einstein statistics displayed by certain atoms.

Electrons too are indistinguishable from each other, but they happen to obey yet a third way of counting, different from the ordinary classical version as well as from Bose's. Electrons behave in a way opposite from that of photons. Where photons tend to crowd together, electrons avoid each other. If the two urns are replaced by energy states in an atom or by opposite directions of spin, a quantum mechanical rule called the *exclusion principle* forbids two electrons to occupy the same one. Thus, the distributions (**, 0) and (0, **) are strictly forbidden, leaving (*, *) as the only option. If this strange rule were magically suspended all of a sudden, all electrons in all atoms would fall down into the lowest available energy state, distinctions among chemicals would vanish, and matter would collapse.

Two simple changes in counting modify the underlying probabilities, which in turn determine the quantum statistics of particles and result in profound consequences for the behavior of matter and radiation. In fact, the consequences are more than profound—they are existential. Without Bose-Einstein statistics or the exclusion principle, the world we know would not exist.

Trying to sort elementary particles the way you sort marbles runs afoul of the problem of inappropriate categories, which we encountered in wave/particle duality and in the notion of a spinning point, all over again. Elementary particles are not trained in human common sense.

Theoretical and experimental surprises such as the cube factory and particle statistics should have raised caution flags when quantum physicists first invoked probability but they didn't. Part of the reason for this failure to think things through more thoroughly may have been physicists' suspicion, verging on disdain, of philosophy. In fact, probability is not only a common, everyday concept that even children use but has also been a subject of debate among scholars for centuries. In any case, for whatever reasons, when quantum physicists finally reached the point where theory meets experiment they lowered their guard, abdicated their critical faculties, and unthinkingly went along with the prevailing definition of probability as "favorable cases divided by all cases."

Because it is based on counting the occurrences of events, this interpretation of the meaning of probability is called *frequentist probability.* It was developed into a rigorous mathematical discipline from the middle of the nineteenth to the first half of the twentieth century and has been taught in schools as a self-evident truth. By its definition as a ratio of numbers, which are accessible to observation, frequentist probability assumes an air of objectivity. The 50 percent probability of a coin toss has the appearance of a real, intrinsic property of the coin, a measurable attribute similar to mass and size.

But even confirmed frequentists don't go quite that far. They claim objective character only for a probability that is derived from a series of coin tosses, not from an examination of the coin or the toss. Their definition of probability must be dug out of statements such as this: "In a large number of fair tosses of a balanced coin the number of heads is about 50 percent, so the probability of throwing heads is approximately 1/2." But mathematicians are not satisfied with the vague words *large, about,* and *approximately.* So they imagine an infinite series of tosses instead. With that change the number of heads reaches exactly 50 percent and the probability 1/2. Unfortunately, the definition also loses its objectivity—it becomes hypothetical and experimentally unverifiable.

Another problem with the frequentist formulation is the word *fair.* It is necessary to assume that the coin is perfectly symmetrical and the manner of tossing absolutely identical for each repetition. But in the real world, symmetrical coins and unbiased tossing mechanisms don't exist. Actually, that's a good thing. If every toss of every coin were identically reproduced in every detail, as we are asked to assume, the outcome would always be the same—at least in a Newtonian, classical, deterministic world. There wouldn't be random sequences of heads and tails, and coin tossing would not be subject to probability theory. So real experiments deal with limited information about the coin and the toss—limited enough to allow some variation but not so limited as to prevent law-like statistical regularities from showing up.

Formal mathematical probability theorists distance themselves from such worries and simply assume exact values for probabilities (such as 1/6 for the hypothetical throw of a die) and infinite runs as primitive axioms—leaving real-world applications to gamblers, pollsters, medical statisticians, and physicists. Mathematicians are aloof from the messy complications of the real world. Secure in the knowledge that a penny will never be tossed an infinite number of times, mathematicians sharpen their definitions and axioms and then prove their rigorous theorems about perfect pennies, unbiased tosses, and infinite patience. Physicists don't have that luxury.

The most consequential principle of the frequentist interpretation of probability is also the one that succeeds most effectively in separating mathematics from real-world experience. It asserts that probability applies to multiple trials but claims nothing about single cases or individual events. For frequentists, "single-case probability" is as meaningless as the concept of "difference" applied to a single number or "attraction" to a solitary particle.

Failure to understand this restriction is related to the *gambler's fallacy,* a bugaboo of school teachers. It is the mistaken belief that after a coin has shown heads 100 times, the chances of tails must surely be higher than 50 percent because a run of 101 heads is ridiculously improbable. In particular, the gambler's fallacy implies that past results for tosses of coins, rolls of dice, deals of cards, and spins of roulette wheels—the long runs of results that define the very notion of probability—predict nothing at all about the next turn. This rule is drummed into the heads of schoolchildren as received wisdom.

Frequentist probability is useful for physicists who deal with multiple trials of carefully controlled experiments, but it rules out the relevance of probability for the single-case probabilities we encounter in our everyday lives. In the context of frequentist probability, statements like "the probability of rain this afternoon is 30 percent," "this milk is probably spoiled," and "she probably loves me," as well as President Obama's reported estimate of the fifty-five to forty-five odds that Osama bin Laden would be found, are all meaningless.

A story serves to underline the gulf between formal probability theory and the way we actually use probability to inform our experiences. Accompanied by a friend, you enter an auditorium where a gambler on stage is tossing a coin and invites you to join in. "I'll bet you a dollar it's heads," he says. "Heads, you pay me a dollar, tails I pay you a dollar. Simple as that!" Confident of avoiding the gambler's fallacy and feeling adventurous, you decide to try your luck. But just as you are about to open your mouth, your friend whispers in your ear: "The last hundred times he did that, he threw heads!"

The question is: What do you do next? Please don't change the story into a sterile textbook problem by asking me: Is the coin fair? Is the friend misinformed? Is the gambler honest? Take the scenario at face value and consider what would actually happen. Please try your best to imagine it as a real-world experience in all its ambiguity and uncertainty. For me, the answer is clear: I would succumb to the discredited gambler's fallacy that allows past events to affect the odds, repudiate frequentist probability theory, and rely instead on my instincts. Even though a hundred heads in a row could theoretically fall by chance and should not affect the next throw, I would not make the bet.

A statistician might defend his theory by claiming that *if* the coin were really fair, the toss really unbiased, and both the gambler and the friend really honest— then I should take the bet. Fair enough, but how am I to know? Without further evidence I would not risk even a dollar. Would you?

What would convince me that the coin is fair? If I, or someone I trust, flipped it a hundred times and it came up heads about half the time in seemingly random order, I would agree with reasonable people that it is indeed fair, at least for all practical purposes. But the reasoning I would have to apply to come to this conclusion is not nearly as straightforward as it seems.

The mathematical physicist Marcus Appleby, who was an early sympathizer with the QBist interpretation of quantum mechanics, illustrated this point with a vivid parable.[1] Imagine, he suggested, that Alice spins a (European) roulette wheel with its thirty-seven numbers once, obtains the number eleven, and concludes that the wheel is fair. Her argument is surely invalid, and any right-thinking person should dismiss it. The result of one spin can't possibly imply anything about the fairness of the wheel. Now imagine a different scenario in which Bob tosses a coin a hundred times, obtains a sequence of heads and tails that, upon examination, consists of about fifty heads and fifty tails in seemingly random order, and concludes that the coin is fair.

If Bob is relying on the observed facts *and nothing else,* his argument is no better than Alice's. From the point of view of mathematical probability theory, a sequence of one hundred coin tosses is equivalent to one spin of a huge roulette wheel with 2^{100} sectors— each labeled with a different sequence of a hundred heads and tails. (If it were designed for a marble-sized ball, this monster machine would not quite fit into the finite volume of the observable universe.) One of the sectors is labeled with precisely the sequence Bob obtained with his coin. So with a single spin of his fabulous wheel, he obtains a result from which he argues that all the other sequences are equally probable and that the wheel, and equivalently the coin, is therefore fair. In spite of the gigantic difference in scale, Bob's argument is as faulty as Alice's.

Appleby invented this tale in order to illustrate a disturbing inconsistency in the frequentist concept of probability. The definition of the probability does not, strictly speaking, exist for a single event. Favorable cases divided by all cases is, instead, a property of the ensemble of a number of repetitions of the event, whether that number is finite or infinite. And yet, as the roulette story will show, the concept of probability

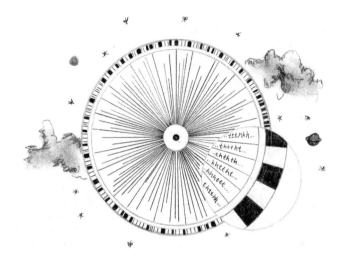

applied to a single event, a so-called single-case probability, is used tacitly by frequentists, even though they cannot define it.

In order to claim that his coin is fair, Bob must in fact reject the roulette analogy and make an argument that rests on unspoken assumptions. He must assume that his one hundred coin tosses are independent and that the probability of heads is the same for every toss. But even that is insufficient. If he makes those assumptions and uses $1/2$ as the numerical value of the probability of tossing heads, he obtains the minuscule probability of $(1/2)^{100}$ for getting the particular sequence he observed. (The number $(1/2)^{100}$ is unimaginably tiny. It is represented by the length of a meterstick that has been cut in half a hundred times over, and purely coincidentally, it's just a little bigger than Planck's constant in metric units.) Unfortunately, there is nothing Bob can do with this infinitesimal probability. It's just like Alice's probability of 1/37 for getting her number eleven and implies nothing at all about fairness. In particular, even an unfair coin could have yielded the very sequence of heads and tails that Bob observed. Bob must delve deeper into the theory and instead of assuming the famous 1/2 for throwing heads with his coin, consider other probabilities as well. Assuming values like 0.7 or 0.2, implying bias for or against heads, respectively, he must repeat his calculation of the probability for the particular sequence he observed. Only now, after this chain of assumptions and calculations, he arrives at a useful result: the probability he calculates for his observed sequence, though tiny, is much larger when he assumes probability 0.5 than what he obtains on the assumption that the coin is biased. Here, finally, is a mathematical answer to the question: Is the coin fair? Yes, because probability 1/2 is quantitatively the most probable assumption.

Notice what Bob was forced to do. Over and over again he referred to probabilities for single, isolated tosses of a penny—to single-case probabilities. First, he had to assume that this probability is the same for each throw—a statement that only makes sense if probability is defined for a single throw. Then he had to assign actual numerical values to that single-case probability in order to find the one that yielded the most likely probability for the entire sequence. Only when that special value turned out to be near 0.5 was he able to claim that his coin is fair.

Marcus Appleby concluded that frequentist probability is not really based exclusively on large sequences of trials, finite or infinite. To be consistent it has to admit single-case probabilities as the fundamental elements, the "atoms" as it were, of probability theory. Frequentism, in short, is inconsistent.

At the end of his paper, Appleby thanked Chris Fuchs, the cocreator of QBism, for making him "see the importance of these questions." This acknowledgment hints at the uphill battle facing QBists. Most of my colleagues in the physics community are blissfully unaware of the problems with the concept of probability. They fail to appreciate the importance of these questions.

10

Probability according to
the Reverend Bayes

Quantum Bayesianism—QBism—is based on an interpretation of probability named for Reverend Thomas Bayes (1701–1761), a Presbyterian minister who was also an able mathematician and statistician. His fame rests on a single paper, published after his death, in which he introduced a special case of a more general result now called *Bayes' law* (also known as Bayes' theorem, rule, formula, or equation).[1] Bayes' law is the lynchpin of *Bayesian probability theory,* a thriving enterprise initiated by, among others, the astronomer and mathematician Pierre-Simon Laplace (1749–1827) and developed by succeeding generations of scholars.

For a century after Laplace, the theory of probability and statistics continued in the Bayesian tradition. Then a number of mathematicians, including John Venn (1834–1923), famous for the eponymous diagram, introduced the frequentist definition in terms of repeated trials in an attempt to make probability "more objective." The seductively simple formula "favorable cases divided by all cases" ended up dominating school teaching. Physicists too adopted frequentism because laboratory experiments in physics are in principle simple, repeatable, and quantifiable. Other sciences, especially fields such as biology, psychology, economics, and medical science, where uncertainties are considerable and unambiguous experiments difficult to come by, struggled to stay connected to the imaginary world of balanced coins and infinite repetitions of experiments. By the middle of the twentieth century, the pendulum of common practice started to swing back to the older Bayesian point of view as an alternative to frequentism. Even astronomers and experimental physicists, drowning in torrents of data requiring statistical analysis, began to rediscover it.[2] Finally, in the beginning of the current millennium, this trend caught up with quantum physics as well, and QBism was born.

Among mathematicians, statisticians, and philosophers of mathematics, the notion of Bayesian probability has been analyzed and dissected and reassembled, resulting in an astonishing number of variations and refinements now available. QBism is based on a version of Bayesianism dubbed "personalist" and "subjective." In this book, that's the only variant I will consider.

Probability is a measure of the likelihood that an event will occur. In ordinary conversation estimates of likelihood are couched in phrases such as "impossible, unlikely, maybe, hard to say, possible, fairly likely, practically certain, certain, there is no doubt whatsoever," but for scientific purposes it is desirable to assign numerical values to probabilities. For simple, idealized situations, such as coin tosses and electrons shot at targets, which can be carried out under controlled conditions, frequentist probability does that job. But in the interest of logical consistency, as we have seen, as well as for practical purposes, a definition of probability that applies to unique events is required. Frequentism cannot oblige.

Bayesianism removes the location where probability resides from the external material world and places it instead in the mind of a person, called an *agent.* In this context an agent (from the Latin *agens* for "doing") is not a representative of other people but someone who is capable of making decisions and performing actions. Bayesian probability is a measure of an agent's personal *degree of belief* that an event will occur or that a proposition is true. The word *agent* imbues the definition with the possibility of real consequences—private musings that do not affect the world in any way are of no interest to science. A "belief" is personal and subjective. It is formed as a result of many diverse influences—only the agent in question knows exactly what they are. Bayesians do not presume to delve into the sources of an agent's beliefs or to judge them.

But Bayesians do want to quantify "degrees" of belief. How do you measure the intensity of a belief? You can't, unless it results in some externally discernible action. The clever device used to convert qualitative estimates into numbers is a formalized version of betting. An agent is cast in the role of a bettor. The amount she is willing to gamble in an imagined monetary bet—regardless of how she comes up with her decision—is used to define her estimate of the probability that an event will occur. Probability theory thus returns to its ancient roots in gambling and games of chance.

In order to standardize the betting procedure and to make sure that probabilities so measured turn out to be real numbers between 0 and 1 (or equivalent percentages), the Bayesian definition is developed as follows: A standard contract between the betting parties takes the form of a coupon bearing the words, "If the event E occurs, the seller of this coupon will pay the buyer one dollar." Once the bettors agree on the precise description of the event E, they buy and sell such coupons among each other. If a buyer thinks that the event is sure to occur—for instance, that the sun will rise tomorrow—she assigns a probability 1 to the event. She will then be willing to pay any price up to one dollar for the coupon. (Paying a full dollar would give her no chance of winning anything—a silly bet.) On the other hand, if she thinks event E will not occur—for example, that her coffee cup will levitate toward the ceiling when she releases it—she will assign probability 0 and will pay nothing for a coupon.

The procedure can be extended to events that are neither certain nor impossible. In the case of a coin flip, for example, the agent has learned in school and from her own experience that the probability of getting heads (the event E in this case) is supposed to be 1/2, so she will pay up to fifty cents for a coupon. Then the coin is flipped. If it's heads, she gets a dollar back and thus nets fifty cents or more. If it's tails, she forfeits her fifty cents or less—a fair bet.

In general, the formal Bayesian definition of a probability, unscientific as it may sound, is the following: An agent's assignment of probability p for the occurrence of an event E means that the agent is willing to pay any amount up to p dollars for a coupon worth a dollar if E happens. Conversely, the agent is willing to sell the coupon for any amount from p dollars and up.

The probability thus defined turns out to be a real number between (and including) 0 and 1, just like the frequentist probability. But behind their outward similarity, the two definitions differ radically from each other. For people brought up in one tradition, it's not easy to switch to a completely new point of view. Unlike a new toothbrush, a novel understanding of probability can't be substituted for the old model overnight. For this reason alone, QBism will not take the physics community by storm, but there is no game stopper on the horizon; no simple objection that rules it out summarily. Bayesian probability has proved its worth as a sound and effective tool in much of the scientific and technological world—QBism extends its range of uses to quantum mechanics.

Physicists, including myself, are usually taken aback when they first encounter Bayesian probability. Talking about "degrees of belief" seems so utterly alien to the customary vocabulary of physics. Physicists feel that the "great laws of nature" should have no truck with subjectivity or the beliefs of individual agents. But the alternative, frequentism, has a frustrating way of turning away from the real world and turning instead into sterile, academic textbook talk. By rejecting the gambler's fallacy—that probability can predict something observable about a single-case event—frequentism condemns itself to irrelevance when it comes to making decisions for future action. If the forecast of 70 percent probability of rain this afternoon said *nothing* about what will actually happen, how would that forecast help me to decide whether to take my umbrella when I leave the house? But in fact, the forecast *does* mean something: I interpret the 70 percent prediction as my "degree of belief" about what to expect this afternoon, and of course it influences my decision—just as President Obama's assessment that Osama bin Laden would be found at home influenced his much weightier decision to order the raid on the al Qaeda leader.

If physics is regarded as an epic human adventure rather than a collection of dead facts, then it too requires a constant stream of decisions, and they in turn are based on degrees of belief. Every evaluation of data, every launch of a new calculation, every design of an experiment, every debate and every conclusion, indeed every step of the journey involves decisions among multiple options. And probability estimates of single-case events inform all of them.

In addition to decisions, there are revisions. No feature of Bayesian probability distinguishes it more clearly from frequentism than the possibility of change. Personal degrees of belief change, and therefore probability assignments for events change too. Frequentist probability, modelled as it is on coin tossing, is fixed in stone once it is defined, or perhaps in silver or in copper, but Bayesian probability, which resides in the human mind, can change in midcourse. And this malleability is precisely where the story of Bayesian probability started in the first place. Bayes' law is a mathematical prescription for changing a probability when some new evidence is acquired and modifies the original degree of belief. (Remember how I changed my mind about the gambler in the auditorium.)

Bayes' law answers the following question: Suppose you know or assume the value of the probability that some specific event will occur. Then you come across a new, relevant piece of information, such as an experimental result or an unexpected news item. How does the new information change your probability estimate?

The value of Bayes' law lies in its mathematical rigor. Probabilities are beliefs, and beliefs, as opposed to facts, are malleable. But how do the probabilities and the new information fit together to produce an updated probability? That procedure is a mathematical result as straightforward and as indisputable as the Pythagorean theorem.

An example illustrates the law. Suppose there is a type of cancer with a well-known incidence of 0.5 percent in the general population, meaning that one person in two hundred is afflicted. Suppose further that a new blood test for the disease has been developed and found to be 99 percent reliable—only 1 percent of test results are incorrect. Your doctor suspects you may have the disease, takes your blood sample, and sends it off for analysis. Several days later, to your horror, he calls to tell you that your test result has come back positive.

What is the probability that you actually do have cancer? How much should you worry? Seeing that the test is so reliable, should you assume the worst?

Should you inform your friends and family? Should you get a second opinion? How can you temper your growing panic with a reasonable assessment of your chances? Is there a glimmer of hope in the realization that you might not really have cancer—that the test has produced an erroneous result, a so-called false positive?

Bayes' law provides an orderly way to think about these questions. It is a relationship among four different probabilities, which can all be expressed as numbers between 0 and 1 or as percentages. Let's represent the new information that your test was positive by a plus sign and the event in question, the finding that you actually have cancer, by a frowny. Then $p(+ \rightarrow \otimes)$ represents the numerical answer to the question: What is your degree of belief in the statement "The positive test implies that you really have cancer"? This is the number you should be looking for—the odds on which to base your feelings.

The second component of Bayes' law is the probability that if everyone in the general population were tested, the test would come out positive— sick and healthy people included. Let's call that $p(+)$. Third, we need $p(\otimes)$, the actual probability that you have cancer, before you have even had the test. This is just the incidence of cancer in the general population with which the story began, namely 0.5 percent.

The fourth number, as Bayes realized, is the crux of the calculation. It is represented by $p(\otimes \to +)$ and stands for the probability that if you knew for sure that you have cancer, your test would come out positive. As my symbol suggests, it is, in a way, an inverse probability, answering an inverse question. Not "If I get a positive test, what are my chances of having cancer?" but "If I have cancer, what are my chances of getting a positive test?" The careless confusion of these two questions causes much mischief! They differ as fundamentally as the statements "Most criminals are male" and "Most males are criminals."

Now the machinery is set up. Bayes' law is the simple equation

$$p(+) \times p(+ \to \otimes) = p(\otimes) \times p(\otimes \to +).$$

Intuitively, it is easy to grasp. Written in terms of percentages rather than numbers, it expresses an obvious fact. Out of a total population, you can select all the people who have tested positive $p(+)$, and *from among them* choose only those who have cancer $p(+ \to \otimes)$. Alternatively, you can proceed the other way around and first pick those who have cancer $p(\otimes)$, and *from among them* choose only those who tested positive $p(\otimes \to +)$. In both cases you will end up with the same group of people—those who tested positive and also have cancer.

Let's do the numbers.

The probability of having the disease is $p(\odot) = 0.5$ percent. The second term on the right-hand side, the inverse probability, estimates the likelihood of getting a positive test result if it is assumed that you have cancer. Since the test is so good, you may assume to a good approximation that $p(\odot \rightarrow +) \approx 100$ percent. This is the number that caused your anxiety when your doctor called with the bad news. Knowing that the test is almost 100 percent accurate, most people intuitively feel that the positive result almost surely implies a definite diagnosis of cancer. But they are wrong!

The trickiest ingredient of the formula is $p(+)$, which measures the probability of finding a positive test result in the general population. Well, 0.5 percent of the population does have the disease, which the test will most likely pick up. But another 1 percent of healthy people (who make up the overwhelming majority of the entire population) will unfortunately get an erroneous positive result—a false positive—so the total portion of people who test positive is $p(+) \approx 1.5$ percent.

Put it all together and divide both sides of the equation by $p(+)$. The probability that you're ill, assuming a positive test, is $p(+ \to \otimes) \approx 0.5$ percent $\times 100$ percent$/$ 1.5 percent $= 100$ percent$/3 \approx 33$ percent. (Notice that two of the three percents in the second expression cancel out). The final result produced by Bayes' law is a mere *one chance in three* that you actually have cancer. It is a reasonable compromise between the national cancer statistics, which give you a 0.5 percent probability of having cancer, and the test by itself, which erroneously suggested nearly 100 percent. What a relief! A repeat test is urgently advisable! Since it is unlikely that you will accidentally fall into the false-positive category twice, repeated testing will cut the uncertainty way down—for better or for worse.

The oddly shaped slices of a pie chart for a total population of ten thousand depict actual numbers from which you can check the (approximate) percentages. The sections labeled 49 and 99 in the diagram together comprise all those who tested positive. Since you are in one of these two categories but you don't know which one, your chances of having cancer are about one in three—as predicted by Bayes' law.

In a more general context, the $+$ sign can be replaced by I, meaning new information, and the frowny by E, meaning an event. With those substitutions and again dividing both sides by $p(I)$, Bayes' law takes its conventional form

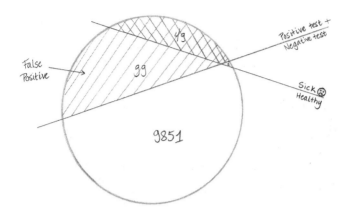

$$p(I \rightarrow E) = p(E) \times p(E \rightarrow I)/p(I).$$

Two of the entries constitute the beef and cheese of the burger, as it were, while the other two make up the bun. The first term on the right-hand side, $p(E)$, is the unadorned probability for the occurrence of event *E before* the new information *I* has been brought to bear. For that reason $p(E)$ is called the *prior probability,* or simply the *prior.* Sometimes it is an uninformed guess made just to get things started, with the anticipation that the repeated application of Bayes' law will improve it. The left-hand side, $p(I \rightarrow E)$, is the new (or *posterior*) probability estimate for the same event *E, updated* by the acquisition of the new information *I.* The other two entries constitute the technical apparatus for effecting the revision. The idea of updating a prior by means of this simple rule is the essence of the Bayesian probability interpretation.[3]

In the cancer example, before the doctor's phone call your estimate of the probability of being ill—the prior—was 0.5 percent. After the doctor's call, your intuitive fear that it had risen to nearly 100 percent was wrong. Bayes' law shows that your estimate should be updated to 33 percent instead.

Bayes' law derives its power from its ability to combine information from very different sources—a feat of integration that is more difficult using frequentist methods, which are better adapted to combining homogeneous data sets. In our example the prior came from large statistical studies in the general population, while the accuracy of the cancer test was presumably measured in controlled clinical studies. Not only numerical data enter into Bayesian calculations. Even history and intuition can help an agent choose a prior and then update it. The example of the gambler in the auditorium, who had allegedly thrown heads a hundred times in a row, underscores the real-life usefulness of allowing added information and fresh hypotheses to affect probabilities, provided they are defined as degrees of belief.

Versatility, generality, and logical consistency recommend Bayesianism over its frequentist rival as the primary interpretation of probability. In climate science, which makes predictions about the manifestly unique atmosphere of the earth and brings together evidence from a vast variety of diverse sources, Bayesian probability theory is often the mathematical technique of choice. Other disciplines, including social science, biology, medicine, and engineering, all use it to advantage. In simple cases the frequentist formula "favorable cases divided by all cases" can determine probabilities numerically, but the Bayesian definition still furnishes their real meaning. The measurement of the area of an oddly shaped sheet of paper illustrates the fundamental difference between a *determination* and a *definition*. Even though the area can be conveniently determined by dividing the paper's weight (in grams) by its density (in grams per square meter), the meaning of the word *area* remains strictly geometrical, with no reference to weight or density.

What happens when Bayesianism meets quantum mechanics, which, as we have seen, relies so fundamentally on the concept of probability?

III

Quantum Bayesianism

11

QBism Made Explicit

L ike a broad river that swells by absorbing countless tiny creeks and brooks along its course, science normally advances gradually as it incorporates
a steady trickle of fresh data and novel ideas. The birth
of Quantum Bayesianism, in contrast, resembled the
confluence of two great streams. At the beginning
of the twenty-first century, quantum mechanics, a
seasoned and sophisticated science at the age of
seventy-five, joined Bayesian probability, a recently
rejuvenated branch of mathematics dating from
the eighteenth century, in a powerful confluence of
well-established bodies of knowledge. The creators
of QBism invented neither the Q nor the B, but they
brought them together—with profound implications
not only for quantum mechanics itself but for the scientific worldview in general.

The principal thesis of QBism is simply this:
quantum probabilities are numerical measures of
personal degrees of belief.

If you haven't heard of Bayesian probability before, this proposition seems bizarre. Isn't the whole
point of science the elimination of the *personal* in
favor of the *universal?* Isn't *belief* the very antithesis

of *knowledge* and therefore of science? That's how most physicists react, and that's how I felt when I first stumbled upon the founding document of QBism published in 2002. The paper boldly announced its startling conclusion right up front in the title "Quantum Probabilities as Bayesian Probabilities."[1]

The decision to switch from the frequentist to the Bayesian interpretation of probability is subject to a kind of cost/benefit analysis. On the one hand, it's fair to ask: What do you gain by the move? On the other hand, what are its downsides—what does it cost you to take the leap?

The cost of adopting QBism is not nearly as great as it might appear because Bayesianism has a sound pedigree. The interpretation of probability in terms of personal estimates of betting odds, though disconcerting at first sight to most people, is not only older than frequentism, but it is also used increasingly by scores of scientists and engineers in the most disparate of fields. It has survived for centuries and passed muster in countless significant applications. For all of its lack of familiarity, it is by no means bizarre.

On the plus side of the ledger, QBism offers considerable benefits, the most convincing of which is the solution to the vexing problem of the collapse of the wavefunction. In the conventional version of quantum theory, the immediate cause of the collapse is left entirely unexplained. There is no mathematical description of how it happens in space and time, as there is for every other process in classical physics. How mechanical, electrical, magnetic, optical, acoustic, and thermal disturbances propagate from point to point and influence objects near and far is understood in meticulous mathematical detail. Even the effects of gravity, the bond that binds us in the universe, can be followed with confidence, step by step from here out to the stars and back again, through the ponderous formalism of general relativity. But the collapse of the wavefunction has remained miraculous—an irritating thorn in the body of mathematical physics.

QBism solves the problem with ease and elegance. In any experiment the calculated wavefunction furnishes the prior probabilities for empirical observations that may be made later. Once an observation has been made—the particle has made its mark, the detector has clicked, the direction of spin has been ascertained, or the position or the velocity has been measured—new information becomes available to the agent performing the experiment. With this information the agent updates her probability and her wavefunction—instantaneously and without magic. The collapse sheds its mystery. Bayesian updating describes it and finally makes the missing step explicit.

The way the process works is straightforward. Consider an example: Alice, in New York, picks two playing cards, one black and one red, and tucks them into separate unmarked envelopes, which she seals and then shuffles. To make sure they are indistinguishable, she asks her friend Bob to shuffle them as well. She keeps one in her purse and hands the other one to Bob. Alice then leaves the room and travels to Australia. Before she opens her envelope, her degree of belief that Bob has the red card is 50 percent. But upon arrival, as soon as she looks at her own card she knows what's in Bob's envelope twelve thousand miles away, so she updates her degree of belief to either 100 percent or to 0 percent instantaneously. In the meantime Bob's guess about the color of Alice's card, whatever it may be, remains unaffected by her actions. There is no miracle.

The collapse of the quantum wavefunction follows the same logic, with one crucial difference. In the classical case, there is an unbroken chain of cause and effect from beginning to end. A material object, in the form of a playing card concealed in an envelope, carries a message in Alice's purse. The card acts as a secret messenger—an example of what physicists call a *hidden variable* with a bit value of red or black. In classical physics Alice's ignorance obscures the value, but she could, in principle, access it at any time along her journey by opening her envelope. In quantum mechanics, on the other hand, there are no cards in envelopes, no objective mechanisms carrying secret messages, and no hidden variables. There is no way, even in principle, to figure out where an electron is, or how fast it's going, or which way its spin is pointing between the time it is fired off and the time it is detected. That there are, in fact, no hidden variables is a claim that can be and has been tested experimentally, and we'll get to that.

When I began to understand QBism and realized that by simply switching to a better definition of probability I could finally stop puzzling over the meaning of the collapse of the wavefunction, I felt a sense of liberation bordering on exhilaration. "Of course," I said to myself, "that's how it works!" It was a delicious feeling of unexpected and undeserved enlightenment—my private eureka! moment.

As if the explanation of wavefunction collapse as a simple updating of a probability were not enough, QBism accomplishes another equally significant clarification. In 1961, just as I was beginning my career, the quantum pioneer Eugene Wigner (1902–1995) pointed out a fundamental ambiguity known as the *paradox of Wigner's friend,* which could equally well be called, "Whose wavefunction is it, anyhow?" Wigner and a friend are conducting a quantum mechanical experiment together. They agree that the system they are observing, say, an electron spin, is described by a *qubit* wavefunction in a superposition of the two possible orientations labeled *up* and *down.* The experiment is performed, and the counter records the outcome. The friend reads the counter while Wigner, with his back turned to the apparatus, waits until he knows that the experiment is over. The friend learns that the wavefunction has collapsed to the *up* outcome. Wigner, on the other hand, knows that a measurement has taken place but doesn't know its result. The wavefunction he assigns is a superposition of two possible outcomes, as before, but he now associates each pole of the electron's *qubit* with a definite reading of the counter and with his friend's knowledge of that reading—a knowledge that Wigner does not share.

So who's right? Has the *qubit* collapsed, or is it still a superposition? As long as the wavefunction is regarded as a real thing or as a description of a real process, the question is no more easily resolved than Bishop Berkeley's infamous question about the tree in the forest: When a tree falls in the forest and nobody hears it, does it make a sound? The answer has been debated for three centuries and still inspires controversy. Einstein, who thought things through for himself instead of relying on ancient authorities, phrased the same question in different terms. His colleague Ernst Pascual Jordan reminisced: "We often discussed his notions on objective reality. I recall that during one walk Einstein suddenly stopped, turned to me and asked whether I really believed that the moon exists only when I look at it."[2] The problem of Wigner's friend—whose wavefunction and whose probability assignment is right?—turns on the meaning of the word *probability* and is as controversial as Berkeley's question.

For the QBist there is no problem: Wigner and his friend are both right. Each assigns a wavefunction reflecting the information available to them, and since their respective compilations of information differ, their wavefunctions differ too. As soon as Wigner looks at the counter himself or hears the result from his friend, he updates his wavefunction with the new information, and the two will agree once more—on a collapsed wavefunction.

The problem of Wigner's friend arose when the question was posed: Who's right? In other words, what *is* the correct wavefunction of the electron? According to QBism, there is no unique wavefunction. Wavefunctions are not tethered to electrons and carried along like haloes hovering over the heads of saints—they are assigned by an agent and depend on the total information available to the agent. They are malleable and subjective. In short, wavefunctions and quantum probabilities are Bayesian.

This terse statement—the QBist manifesto, if you will—is short enough to fit on a T-shirt, but it brings with it a new way to think about the world.

12

QBism Saves Schrödinger's Cat

Schrödinger's cat is probably the most famous feline in the world, but not all physicists are fond of it. I once attended a lecture by Stephen Hawking in which he exclaimed in the mechanical cadence of his voice synthesizer: "When I hear someone mention Schrödinger's cat I reach for my gun!"[1] The QBist pioneer Chris Fuchs also dislikes the animal and told me that he has always preferred to worry about Wigner's friend instead. The cat is a victim of its own fame. Popular culture has encrusted its story with so much misunderstanding, mockery, and outright nonsense that most physicists try to avoid it. But since, at the advanced age of eighty, it's still effective in making a point, I revive it yet once more.

Here's the setup: A living cat is shut up in a box together with a Rube Goldberg arrangement consisting of a Geiger counter, an atom freshly rendered radioactive by neutron bombardment, a hammer, and a vial of poison gas. When the atom decays, as it must eventually do, the Geiger counter clicks and emits an electrical signal, which triggers the hammer, which smashes the vial, which releases the gas, which kills the cat. Instantly and painlessly.

The first question is: How does a quantum physicist describe this experiment? A radioactive atom is associated with a wavefunction represented by a *qubit* whose north pole, labeled 0, represents *intact* while the south pole, labeled 1, represents *decayed*. The probability inferred from the wavefunction drops smoothly from 0 toward 1 at a well-known, ever-diminishing rate. After a time interval that defines the half-life of the atom, the *qubit* has reached the equator, where it is a blend of 50 percent intact and 50 percent decayed. If you observed the atom at that moment, you would have a fifty-fifty chance of finding it decayed.

It is important to note that according to the conventional interpretation of quantum mechanics, which prevailed when Schrödinger invented his cat, the value of a *qubit* is (except at the poles) a blend of "0 *and* 1." It is not "0 *or* 1." Young's classic double-slit experiment displays the difference most emphatically. In order for interference to occur, the light wave must pass through *both* slits, not one *or* the other. By the same token, a point on the *qubit* sphere represents not an alternative but a superposition of both possible outcomes of the quantum event in question. Quantum interference effects are as real and observable as the colors of soap bubbles, and the only way we know how to describe them is using "both . . . and" superpositions.

Thus far, all this is conventional quantum mechanics and undisputed. Countless experiments have demonstrated that it is the correct way to describe a radioactive atom. The trouble starts when you make inferences from the atom to the cat itself. What is the state of the cat after one half-life of the atom, provided you have not opened the box? The fates of the cat and the atom are intimately connected—*entangled* is the evocative term introduced in English by Schrödinger himself. An intact atom implies a living cat; a decayed atom implies a dead cat. It seems to follow that since the atom's wavefunction is unquestionably in a superposition so is the cat: it is both alive *and* dead. As soon as you open the box, the paradox evaporates: pussy is either alive or dead as common sense dictates. But while the box is still closed—what are we to make of the weird claim that the cat is dead and alive at the same time?

Schrödinger concocted the story in order to bring quantum weirdness from the obscure realm of individual atoms and their wavefunctions up into the daylight of human experience. He sought to dramatize the difference between the two domains. Much of the impetus for developing the alternative interpretations of quantum mechanics that have been invented over the last ninety years sprang from mathematical elaborations of the cat scenario.

QBism deals with the story as effortlessly as it disposes of the miracle of wavefunction collapse and the paradox of Wigner's friend. The map is not the territory! The wavefunction of the atom is not a description of the atom. The *qubit* describing the atom is a summary of a particular agent's belief about the betting odds for a future observation—nothing more and nothing less. The state of the atom, *before* it is observed, is defined mathematically but not in the terms we use after we actually observe it. According to QBism, the state of an unobserved atom, or a quantum coin, or a cat for that matter, has no bit value at all. A point on the equator of a *qubit* ball is not a symbol for anything in the real world—it merely represents an abstract mathematical formula that gives the odds for a future observation: 0 or 1, intact or decayed, dead or alive.

Claiming that the cat is dead *and* alive is as senseless as claiming that the outcome of a coin toss is both heads and tails while the coin is still tumbling through the air or that a horse has won and lost before the race is run. Probability theory summarizes the state of the spinning coin by assigning a probability of 1/2 that it will be heads. The tote board at the racetrack lists the odds for the horse winning. In the same way, QBism refuses to describe the cat's condition before the box is opened and rescues it from being described as hovering in a limbo of living death.

A memorable way to describe this conclusion was formulated in 1978, long before the advent of QBism, by the theoretical physicist Asher Peres (1934–2005). He noticed that stories like that of the cat involve a "What if?" question: "What if we could look at the cat while the box is still closed?" Peres concluded that quantum mechanics does not allow "What if?" questions and coined the catchy slogan "Unperformed experiments have no results." Classical physics, of course, permits imagining what's in the box before it is opened. The result of this classical thought experiment is that the cat is dead *or* alive. In quantum mechanics, however, there is a well-defined way to describe a system that is in one of two possible states, state 0 *or* state 1. The mathematical tool for such a description is a classical bit—the universal toggle switch of information technology. But the bit is not available as a possible wavefunction of a radioactive atom. The *qubit,* which replaces the bit in the context of quantum mechanics, has no bit value at all until a measurement has been performed. Describing atoms by means of bits instead of *qubits* leads to blatant conflicts with experiments.

Peres's formulation is profoundly QBist in spirit. If the wavefunction, as QBism maintains, says nothing about an atom or any other quantum mechanical object except for the odds for future experimental outcomes, then an agent won't even be tempted to speculate prematurely about the state of the atom or of the cat. The unperformed experiment of looking in the box before it is opened has no result at all, not even a speculative one.

The bottom line: According to the QBist interpretation, the entangled wavefunction of the atom and the cat does not imply that the cat is alive and dead. Instead, it tells an agent what she can reasonably expect to find when she opens the box.

13

The Roots of QBism

Though QBism is a twenty-first century innovation, its roots can be traced all the way back to the Greek atomists. Democritus, who lived around 400 BCE, taught that "sweet is by convention, and bitter by convention, hot by convention, cold by convention, color by convention; in truth there are but atoms and the void." People might privately disagree about what to call sweet or bitter, hot or cold, but they have to agree on the presence or absence of particles of matter, provided their senses and their instruments are sharp enough.

Democritus is hailed as the father of atomism on the basis of his declaration. "In truth there are but atoms and the void" sounds authoritative, doesn't it? Reassuring, persuasive, definitive. What might be called the atomist manifesto held physics in thrall for two-and-a-half millennia and became the conventional wisdom taught in school. On page 2 of his classic *Lectures on Physics,* Richard Feynman reaffirmed the atomist manifesto in his own words:

> If, in some cataclysm, all of scientific knowledge were to be destroyed, and only one sentence

passed on to the next generations of creatures, what statement would contain the most information in the fewest words? I believe it is the *atomic hypothesis* (or the atomic *fact,* or whatever you wish to call it) that *all things are made of atoms—little particles that move around in perpetual motion, attracting each other when they are a little distance apart, but repelling upon being squeezed into one another.* (Italics in the original)

Throughout my years in the classroom, I have taught the atomist manifesto according to Democritus and Feynman. Imagine my surprise when I found out that the aphorism attributed to Democritus, for all its enduring influence, is incomplete. It is actually part of a little dialogue that reads in full:[1]

Intellect: "Sweet is by convention, and bitter by convention, hot by convention, cold by convention, color by convention; in truth there are but atoms and the void."

The Senses: "Wretched mind, from us you are taking the evidence by which you would overthrow us? Your victory is your own fall."

The passage is not an unequivocal atomist manifesto but a whimsical caricature of the conflict between two profoundly different ways of perceiving nature. According to the Intellect, science strives to describe the world as it "really" is; the scientific mind aspires to discover the true essence of things. On this view the attention of the scientist is entirely focused on the object in question, be it a tree, a rock, an atom, or an electron. There is no room here for the observer who is describing the object and its behavior. Science aims for objectivity—subjectivity is taboo.

But the Senses object to their exclusion from the description of nature. They remind the Intellect of the obvious fact that everything we know about the universe we learn from sense experiences, either directly or with the aid of instruments. See that tree over there? How do you know what it *really* is? You discover its colors and shape with your eyes, aided by optical devices. You can walk over and touch it to feel the hardness of its wood. You can smell the scent of its blossoms. You can remember what you've learned about it from your own observations and from reading what others have said about it, but between the tree and your mind, which tries to make an accurate map of it, your personal sense experiences always serve as messengers. And if that's the case for trees and rocks, it's also true for electrons and quarks and for matter and space and time.

Realizing this simple fact, the Senses conclude that if the Intellect dismisses their crucial role in science as mere convention, it discards the only evidence it has for finding out what it chooses to call the "truth."

For centuries after Democritus, philosophers and theologians thought deep thoughts and wrote thick treatises about the relationship between reality and our perception of it; between what *is* and what *appears* to be. Physicists, however, ignored those debates. They suppressed the second half of Democritus's fragment, purged their accounts of subjective influences, and constructed what they claimed was a purely objective description of a world without observers. They could get away with this strategy because they rigorously confined their attention to simple, inanimate systems like orbiting planets, falling apples, and inert particles of matter. By asking simple questions, they were successful in discovering simple, seemingly objective answers.

Strict objectivity worked spectacularly well for centuries, but the spell of Democritus was doomed to end, just as he foresaw. "Your victory is your own fall," the Senses warn the Intellect. Einstein's special theory of relativity of 1905 made a conspicuous break with absolute objectivity by smashing Newton's austere, intuitively appealing scaffolding called *absolute space* and *absolute time.* Without that rigid background to define motion, a statement such as "That car is moving at fifty miles an hour" lost its meaning.

Relative to a stationary cop that might be true, but if she is pursuing the car in her cruiser, she will measure a different speed. The observer, or at least the observer's frame of reference, must always be specified in order to make any sense of mechanics. Einstein's crucial clarification was quickly shown to be not just a pedantic quibble but an important insight with dramatic observable consequences. The lofty absolute space and time born of Newton's Intellect were replaced by Einstein's more mundane relative space and time, which yielded much better agreement between theoretical predictions and laboratory measurements. Although the theory of relativity did not explicitly reintroduce observers, their freely chosen frames of reference, at least, have assumed an indispensable role in physics.

Another assault on unvarnished objectivity came with wave/particle duality. An electron is not really a particle or a wave but a hybrid that reveals different properties depending on the questions asked and the apparatus the experimenter freely chooses. When the full-blown quantum theory appeared in 1925–1926, Democritus's astute prediction inched even closer toward fulfillment. With the introduction of the wavefunction, physicists stopped trying to describe electrons, photons, atoms, and nuclei "as they really are." A particle does not really have a speed and a position, but one or the other, depending on how you choose to look at it.

Attention pivoted from the territory to the map as the gaze of physicists shifted away from the real world—which undoubtedly exists out there—to its representation. Separating the thing from its mathematical description was a significant but largely unheralded break quantum mechanics made from its classical parent.

The pioneers of the quantum theory understood this radical implication of their work. Niels Bohr, who did not invent quantum mechanics himself but contributed significantly to its interpretation, wrote in 1929, three years after Schrödinger's introduction of the wavefunction: "In our description of nature the purpose is not to disclose the real essence of the phenomena but only to track down, as far as it is possible, relations between . . . aspects of our experience."[2] The "real essence" corresponds to Democritus's "truth," and "our experiences" refers to our senses. Essences are objective, absolute, and universal; experiences are subjective, relative, and particular to individual agents.

Werner Heisenberg, who invented quantum mechanics with his matrix treatment of the harmonic oscillator, insisted that "the conception of objective reality . . . has thus evaporated into the . . . transparent clarity of mathematics that represents no longer the behavior of particles but rather our knowledge of this behavior."[3] Physics, he believed, is not about this tree or that electron, as Newtonian science had assumed, but about what happens in our minds as a result of observations and experiments concerning the tree and the electron. The phrase "no longer" clearly telegraphs the break he perceived with classical physics.

Erwin Schrödinger himself put it this way in 1931: "One can only help oneself through something like the following emergency decree: Quantum mechanics forbids statements about what really exists—statements about the object. Its statements deal only with the object-subject relation."[4] In other words, quantum mechanics describes what an observer (the subject) experiences while contemplating nature (the object).

Succeeding generations of physicists did not pay much attention to such philosophical caveats. Worries about "essences," "methods of questioning," "emergency decrees," and "object-subject relations" did not particularly concern them. They quickly realized that the new quantum theory, together with rapid improvements in technology, combined into an amazingly robust tool. The understanding of matter at the atomic and nuclear level progressed by leaps and bounds. New quantum devices, such as transistors and lasers, were in turn used to probe ever more deeply into the atom, even as they were turned into consumer goods ranging from computers to cell phones. Quantum mechanics *worked*. A rush of discovery and invention during the second half of the twentieth century largely swept aside philosophical qualms about wave/particle duality, superposition, uncertainty, and wavefunction collapse.

But the weirdness persisted. The crux of the problem, as is often the case in global conflicts, is a boundary dispute. On one side is the familiar world we perceive with our senses and describe in deterministic, Newtonian terms. It is characterized by great laws of nature and, in principle at least, by certainty. On the other side, we find the world of the quantum, a world of uncertainty and of probability. The question is: Where does one territory end and the other begin?

Initially, the answer seemed obvious. Since quantum mechanics was developed for electrons, photons, atoms, and nuclei, the impression arose that quantum phenomena were necessarily confined to the microworld teeming with incredibly small objects in unimaginably large numbers. This error suggested a division of modern physics into four adjacent regions: The very large is governed by general relativity, the very fast by special relativity, and the very small by quantum mechanics. The three modern branches of physics surround the human-scale classical region where Newton reigns.

But that neat scheme failed for two reasons—one practical, the other philosophical. Quantum effects were found in ever-larger systems. The double-slit interference experiment, for example, which started with photons and electrons, was repeated with atoms and even fullerenes, huge molecules composed of sixty or seventy carbon atoms. Will viruses be next? And then cats? More recently, as I mentioned in Chapter 4, a conventional, though tiny, tuning fork was shown to display quantum behavior. On the astronomical front, planet-size neutron stars were discovered to behave like gigantic nuclei. Even the entire universe is thought to have behaved quantum mechanically in its infancy. Clearly, the notion that quantum mechanics applies only to the microworld is simply wrong.

The philosophical objection to confining quantum mechanics to atoms and molecules is even more persuasive. The complaint about tigers and sharks, who rule the grainy deserts and the wavy oceans, applies here as well. There should not be two theories—the classical and the quantum mechanical—with different foundations and only a miraculous, fragile bridge called the "collapse of the wave function" to connect them. There should be only one theory, from which the other can be derived with a simple, compelling argument. Either we live in a classical world and quantum mechanics is a mere approximation or it's the other way around!

The line between quantum land and our land is a fuzzy, disputed border. Heisenberg, after whom it is sometimes called the *Heisenberg cut,* thought of it as the border between a quantum system, such as an atom, which is described by a wavefunction, and the apparatus for observing it, which follows classical rules. He tried to make a virtue of its undefined location by shifting the cut around—treating a cat or a colleague either classically or as a large quantum object at his convenience. This kind of waffling did not impress John Bell (1928–1990), a brash, brilliant physicist whose great claim to fame was to bring the disputes about the meaning of quantum mechanics from theoreticians' offices down into the laboratory, where they could be resolved experimentally. He mocked the cut, calling it a "shifty split" too vague to be a useful concept for serious analysis.

Over the years, the term *Heisenberg cut* has been variously applied to the dividing line in dichotomies such as macroscopic versus microscopic, classical versus quantum, Intellect versus Senses, objective versus subjective, certain versus uncertain, real versus apparent, physical world versus observer, territory versus map. . . . Invariably, the split is blurred, ill-defined, and shifty. Finally, the distinguished Cornell physicist N. David Mermin, a man of my own generation and a fellow convert to QBism, proposed an end to the discussion. He "called the question," to use a parliamentary term, suggesting that so much ink has been spilled on the subject that further debate seems futile. In 2012 he wrote an essay[5] whose subtitle announced his intention of "fixing the shifty split."

(Mermin has a way with words. His term *fixing* implies both repairing and stabilizing.) QBism, Mermin argued, offers a clear and compelling suggestion for locating and defining the split. It is indeed the boundary between what is objective (external, unaffected by thoughts and feelings, existing independently of perception) and what is subjective (internal, perceived, existing in the mind). But in contrast to what previous scholars labeled subjective, that is, existing in the human mind, for QBists the subjective is also strictly personal: it exists in one particular person's mind. The split, according to Mermin, belongs to each agent individually.

Each of us is aware of the difference between the (objective) world and the (subjective) awareness of our own experience. If I am the agent, the objective world is everything outside my mind—including other agents and even my own body. All of that I may, if I choose, treat quantum mechanically and describe by wavefunctions. On the other side of the split are things exclusively personal to me, and those neither I nor anyone else can treat as objects. They are my own experiences and perceptions. They serve as input for the beliefs I hold and the bets I make about future experiences. They are subjective and uniquely personal.

If a layman and a QBist happened upon a closed box containing Schrödinger's cat, the layman would confidently declare: "From past experience I know that the cat is either dead or alive." He would be talking about the cat at that moment. The QBist would be more cautious and say: "I don't know anything about the cat at the moment. But according to my knowledge of quantum mechanics, I believe that if I opened the box right now, the chances are fifty-fifty that I would find it alive." Thus, neither the layman nor the QBist would claim that the cat is both dead *and* alive, but the QBist would be talking about her own beliefs about a future experience, not about the current state of the cat.

Hearing her, Democritus, whose nickname is the Laughing Philosopher, would smile. After more than two millennia, his warning is finally being heard. The Intellect is beginning to respect the Senses.

14

Quantum Weirdness in the Laboratory

In the early days, the persistent conceptual problems of quantum mechanics had a distinctly oth-
erworldly flavor. Since the theory worked so well in
practice and the paradoxes seemed to be related more
to the interpretation of the formalism than to its
content, most physicists felt that they could safely
ignore them. Problems such as the collapse of the wavefunction, Wigner's friend, and Schrödinger's cat belong to the realm of *thought experiments*—theoretical
exercises of such precious refinement that they could
not possibly be repeated in the laboratory. You can't
catch a wavefunction in the act of collapsing or determine the well-being of a cat without looking at it.

But thought experiments should not be summarily
dismissed—in time they often turn real. At the beginning of the twentieth century, for example, Albert Einstein kick-started special and general relativity with
thought experiments that eventually found their way,
in considerably amended form, into the observatory
and the laboratory, with historic consequences. In
1935 he did it again in a paper written with colleagues
Boris Podolsky and Nathan Rosen titled "Can the
Quantum-Mechanical Description of Physical Reality

Be Considered Complete?" The authors (EPR) noticed that if you could actually perform a certain type of atomic experiment and describe it quantum mechanically, you'd come up with weird, contradictory conclusions, which caused Einstein to cast doubt on the theory as it was then understood. The argument, known as the *EPR paradox,* has sparked a vigorous, seemingly unending debate in the small community of philosophers, historians, and physicists concerned with the foundations of physics. After Einstein's death in 1955, his thought experiment began to become real.

Instead of following the historical thread, I will skip the versions of the EPR experiments that were eventually performed and proved Einstein's worries about quantum mechanics to be unfounded.[1] Instead, I'll fast-forward to the present century, illustrating the idea of EPR with a different experimental arrangement that is easier to understand than the original example. Unlike its predecessors it does not depend on the analysis of subtle statistical correlations or on the role of randomness in quantum phenomena but hinges on a single observation that demonstrates the conflict between quantum mechanics and common sense in one decisive blow.

EPR suggested that the interplay of two general assumptions, both of which Einstein considered self-evident, leads to the conclusion that conventional quantum mechanics is wrong, or at least incomplete.

If, on the contrary, quantum mechanics is correct as it stands, you must give up one of those two assumptions. Einstein could not bring himself to abandon either one and was therefore reduced to hoping that quantum mechanics would one day be made whole. Most physicists, including QBists, believe that quantum mechanics is the full and correct theory of the world and are therefore forced to give up one of the two EPR hypotheses.

The two critical hypotheses, which happen to hold for classical physics, are *locality* and *realism*.

Locality is the absence of what Einstein called *spooky* action at a distance. A local theory is one in which signals and other physical effects do not travel with infinite speed. Instead, they propagate through space from point to nearby point in domino fashion, at a speed that cannot exceed the speed of light. Newtonian gravity, with its instantaneous action at a distance, spectacularly violated the principle of locality and was replaced by general relativity, which respects it.

In quantum mechanics, violations of locality seem to occur under two circumstances. The collapse of the wavefunction, as we saw, is a nonlocal process that QBists explain by interpreting probability as belief rather than physical reality. The experiments of the EPR type seemingly violate locality in a related but different way. They purport to show that the measurement of a physical quantity in one place instantaneously, or at least at superluminal speed, influences the outcome of another measurement far away. Magicians call that sort of thing telekinesis—the art of moving objects by the power of thought alone. Einstein called it spooky.

The experimental demonstration of such effects is nevertheless so startling that some physicists believe that the world is, in fact, nonlocal. That the universe is one single interconnected object that trembles when you tickle it far away is a poetic notion, to be sure, but its denial has been a considerably more fruitful approach to understanding the workings of the material universe.

The second supposedly self-evident assumption underlying EPR is more difficult to pin down. By *realism* I am of course referring to scientific rather than literary, artistic, or philosophical realism. But when you look up the 30-page essay on "Scientific Realism" with its bibliography of over 180 items in the authoritative online *Stanford Encyclopedia of Philosophy,* you find the discouraging caveat: "It is perhaps only a slight exaggeration to say that scientific realism is characterized differently by every author who discusses it." Ouch.

Relying once more on Einstein's homespun wisdom, one might try to define realism as the assumption that the moon is there even when nobody looks at it. More generally, it refers to the assumption that objects have physical properties that are unaffected by measurements and observations. One might go further and propose that *real* means unaffected by measurements, observations, and even thoughts and opinions. EPR defined reality thus: "If, without in any way disturbing a system, we can predict with certainty ... the value of a physical quantity, then there exists an *element of reality* corresponding to that quantity."[2]

To see how this assumption works in practice, we might think of an astronomical observation—which surely does not disturb the system. When Galileo discovered the moons of Jupiter, skeptical astronomers believed them to be artifacts—stray reflections or imperfections in the lenses—of the primitive telescopes then in use. In fact, sometimes there were three and at other times four little dots in the sky next to the giant planet, and their positions seemed to shift nightly. But eventually, regularities were established, the disappearances were explained as moons passing in front of or behind the planet, and the predictions of the observed positions of the moons achieved certainty. From then on the moons and their positions in the sky became elements of reality.

To summarize the EPR claim: Quantum mechanics is incompatible with the simultaneous assumption of locality and realism. If, with Einstein, you insist on both of those, you must find fault with quantum mechanics. This is an astonishingly broad claim. Most physical predictions are much more specific and more modest, along the lines of: if this ball is dropped from a height of four feet, it will hit the ground in half a second. And yet, experiments have been performed to prove the EPR claim with its vague, ambiguous, philosophical premises.

I will describe one such experiment in terms of *qubits,* ignoring the substantial instrumental complications it entails. Furthermore, although it was performed with photons, I'll describe it in terms of electrons instead because electrons are material particles and a bit more accessible to our intuition than photons. The appeal of *qubits* is their ability to succinctly describe any two-state quantum system, be it a photon with two possible polarizations or an electron with two orientations of spin along some arbitrary axis.

Before I begin I must introduce a logical device that plays a useful role in the analysis—the notion of transitivity. Transitivity is just common sense. It claims that if Alice and Bob have eyes of the same color and Bob and Charlie do too, then Alice and Charlie must share eye color as well. Equality is transitive: If $A = B$ and $B = C$, then it follows by common sense as well as logic that $A = C$. The transitive relationship needed for the quantum experiment concerns the geometric property of direction. If spins *A* and *B* point in the same direction and if the same goes for *B* and *C,* then *A* and *C* necessarily point in the same direction too. Keep in mind that the spin of an electron can only be measured along one axis at a time, though.

With EPR, *qubits,* locality, realism, and transitivity, the pieces of the puzzle are on the table.

The simplified and idealized experiment I am about to describe was proposed in more realistic terms by Daniel Greenberger, Michael Horne, and Anton Zeilinger (GHZ) in 1989 and performed in 2000. It proceeds in four phases—preparation, measurement, prediction, and analysis.

Preparation

Three electrons are brought into close contact and wrestled into a very special configuration called an *entangled state.* Its spin wavefunction is stitched together from three *qubits* that will be represented by three arrows, corresponding to measurements in the vertical or horizontal directions. The electrons are neither observed nor are their spins measured while they are in each other's vicinity.

After this crucial and technically challenging preliminary step, the electrons fly off to three widely separated spots, where three independent detectors observe their spins. The configuration has been rigged in such a way that when two of the three spins point in the same horizontal direction the third, measured in the vertical direction, points up. On the other hand, if two horizontal spins oppose each other the third vertical spin points down. Abbreviating right, left, up, and down using *R, L, U,* and *D,* the only possible observations are *RRU, LLU, RLD,* and *LRD.* Symbolically, these are represented by $(\rightarrow \rightarrow \uparrow)$, $(\leftarrow \leftarrow \uparrow)$, $(\rightarrow \leftarrow \downarrow)$, and $(\leftarrow \rightarrow \downarrow)$. Since the three electrons are interchangeable, the order of the arrows in each pair of parentheses is irrelevant, so the last two possibilities are actually equivalent.

A mnemonic helps to remember the scheme: If your two index fingers point in the same horizontal direction, they "agree" and one of your thumbs points *up.* If they point in opposite horizontal directions, they "disagree" and the thumb points *down.*

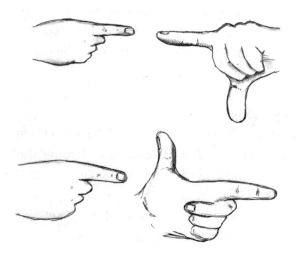

This preparation can be checked over and over again, each time with a new trio of electrons and with two detectors in the horizontal orientation and one in the vertical. It is robust. From any two measurements, the third can be predicted with certainty and would have been called an element of reality by EPR. I call this restriction on possible results the *GHZ rule*. Throughout the entire experiment, the preparation of the three electrons does not vary in any way.

Measurement

After being prepared in this way, the entangled electrons part company, and measurements are performed on their spins with the distant detectors. However, the detectors are now oriented in different directions from what they were when the preparation was being checked. In particular, all three detectors are turned to measure only vertical spins. The first two result in a mixture of *UU, UD, DU,* and *DD.* Only the *UU* events are kept. The others are ignored.

Predictions

What does the third detector find? The first *U* implies that the horizontal spins of electrons two and three, *if they were measured,* would agree. The second *U* implies that horizontal spins one and three, *if they were measured,* would also agree. By the principle of transitivity, as well as common sense, this implies that electrons one and two must agree and therefore that the vertical spin of the third electron is (thumb) *up.*

In short, the classical prediction is that the three detectors should measure *UUU.*

Quantum mechanics, on the other hand, predicts unequivocally that the configuration *UUU* is forbidden and that *UUD* is the only allowed result. This prediction follows directly from the GHZ wavefunction, but I cannot explain it any better than that. What matters is that it was, in fact, experimentally confirmed. Get used to it!

That *UUD* result is a call to arms, loud and clear and undeniable. More impressively than any other single observation, it signals the need for a revolution in thinking.

Analysis
Quantum mechanics has won the contest over common sense—now we must examine the implications for locality and realism, which, according to EPR, cannot both continue as the law of the land.

First, let us insist on realism. A property of an object is real if the object carries it along—if its value preexists any measurement and an observation only reveals, not creates it. Remember the red and black cards in Alice's and Bob's sealed envelopes in Chapter 11—those are real and exist even before the envelopes are opened. Let us assume then that spin directions are also real attributes of an electron. Let us also suppose that, contrary to the laws of quantum mechanics, the values of both vertical and horizontal spin can be assigned to each electron simultaneously, always subject to the GHZ rule (*RRU, LLU, RLD,* and *LRD*).

Under this assumption, the spins are manipulated and preassigned while the electrons are together at the beginning of the experiment. Only two assignments (and their mirror images) obey the required rule. In this symbolic scheme, each pair of arrows refers to the (simultaneous) vertical and horizontal spin values of one electron. (Once more I remind you that quantum mechanics, and in particular an uncertainty principle, forbids the simultaneous measurement of horizontal and vertical spin.) Here are the only allowed configurations:

↑→ ↑→ ↑→ and its mirror image ↑← ↑← ↑←

or

↓→ ↓→ ↑← and its mirror image ↓← ↓← ↑→.

Please check that the four assignments do indeed obey the GHZ rule.

All other assignments fail to obey the rule. For example, can you spot where the rule breaks down for these assignments that include two *up* measurements?

$\uparrow\rightarrow \uparrow\rightarrow \downarrow\rightarrow$ and its mirror image $\uparrow\leftarrow \uparrow\leftarrow \downarrow\leftarrow$

or

$\uparrow\rightarrow \uparrow\leftarrow \downarrow\leftarrow$ and its mirror image $\uparrow\leftarrow \uparrow\rightarrow \downarrow\rightarrow$.

To see in detail how these results come about, start with $\uparrow$ and build up the rest of the configuration, always following the GHZ rule. You very quickly conclude that it is impossible to recover the observed result *UUD* with preassigned spin values. The only way out is to invoke a spooky effect: The two initial measurements *UU* somehow affect the last measurement, at a distance, to force it to be *D,* the result that quantum mechanics correctly predicts. If you insist on realism, locality is violated.

If, on the other hand, you give up realism (as the QBists do), locality can be saved. In that case the electrons interact initially in one locality to produce an entangled trio described by a quantum wavefunction that incorporates the GHZ rule. Since it is not real, the wavefunction does not claim to describe a real state of affairs the way all the little arrows above purport to do. Instead, the wavefunction is a cunning mathematical construct made of *qubits* that correctly predicts the outcomes of the GHZ experiment, in both its preparation and measurement phases.

The GHZ experiment provides a splendid illustration of the maxim "Unperformed experiments have no results." The contradiction between classical and quantum physics only came about when we assumed that in the final phase of the experiment the *horizontal spins* had definite values, even though they were not measured. Peres's cautionary remark forbids the simultaneous assignment of two spin directions to one electron, expressed by symbols such as $\uparrow\rightarrow$.

An alternative way to analyze the GHZ experiment is in terms of hidden variables, which carry concealed messages like red and black playing cards in sealed envelopes. Many of the predictions of quantum mechanics can be accounted for—as Einstein hoped—without sacrificing either locality or realism. If you assume the existence of hitherto undiscovered attributes that carry information, you can reproduce much of quantum mechanics by adjusting the values of these attributes. In the GHZ experiment, for example, this program would succeed all the way through the preparation phase. The GHZ rule can be enforced by regarding the vertical and horizontal spins as hidden variables capable of being assigned simultaneously, even though they cannot be measured simultaneously. In the GHZ experiment, quantum mechanics, locality, and realism happily coexist with hidden variables, provided that two of the detectors are horizontal and the third, vertical.

The crux of GHZ is the ingenious discovery that when all three detectors are vertical, no amount of mental gymnastics—not even the hypothesis of hidden variables—can get around the flat contradiction between quantum mechanics and common sense. Hidden variables, such as cards in sealed envelopes, allow classical physicists to tell an uninterrupted, believable story about what happens between measurements in any experiment. That possibility amounts to the claim that we understand what really happens even if we don't prove it by observation. It amounts to the assumption of realism. Quantum mechanics, however, forces us to abandon such stories. Asher Peres's admonition that unperformed experiments have no outcomes warns of the dire consequences of trying to concoct them.

The GHZ experiment does not prove the correctness of QBism, but QBism, by foregoing realism of the Einstein, Podolsky, and Rosen kind, provides a simple, convincing way to avoid spooky action at a distance.

All Physics Is Local

Quantum mechanics does not include explicit action at a distance. In the GHZ experiment, for example, the wavefunction is made of a combination of three *qubits* to describe three electron spins. Position and time are not even mentioned in this expression, so the distance in the phrase *action at a distance* is irrelevant. In contrast, Newton's venerable law of universal gravitation, which claims that as I move, my attractive force on your body changes simultaneously, is an example of explicit, instantaneous action at a distance. But how you *use* the GHZ wavefunction, what you *do* with it, and how you *interpret* it may lead you to believe in implied action at a distance. We saw that if you insist that the wavefunction is real, you are forced to conclude that the detectors must somehow communicate with each other at a distance—causing outcomes that depend on the results of faraway measurements. How such a spooky effect comes about seems as mysterious to us as gravitational attraction was to Newton.

Einstein's special and general theories of relativity banished explicit (as opposed to implied) action at a distance from physics. Fundamentally, all physics is local, to paraphrase the venerable American maxim that all politics is local.

Richard Feynman invented a clever way to drive this point home. The electrons in an atom are subject to ordinary electrical forces—attraction to the nucleus and repulsion among each other. In a crude, classical theory, those forces are described in exactly the same way Newton described gravity, by action at a distance: unlike charges attract, like charges repel. That's an approximation good enough to derive the atomic wavefunctions of the early quantum theory. Eventually, though, electrical and magnetic interactions themselves were quantized so that not only the electrons but the very forces between them were subjected to the rules of quantum mechanics. The theory that managed to accomplish this task, which was perfected in the middle of the twentieth century and aptly named QED for quantum electrodynamics, combined quantum mechanics with classical electrodynamics to describe the behavior of photons and electrons with awesome precision. In Chapter 8 I mentioned the magnetic strength of an electron as one of its successes.

As this theory was refined to improve its agreement with experiments, its complexity increased rapidly. Eventually, it required reams of dense calculations, with the result that errors inevitably crept in. Feynman, with a keen eye for clever effort-saving tricks, noticed common patterns in the equations that prompted him to develop a suggestive graphical language—a kind of mathematical shorthand for quantum computations. *Feynman diagrams* are so simple that physicists scribble them on paper napkins in restaurants in order to illustrate abstruse points buried deep in the math. At the same time, though, every line and squiggle in a diagram is backed up by a detailed recipe for translating pictures into formulas. Feynman diagrams soon became the universal symbolic language—the lingua franca—for particle physicists throughout the world.

The first Feynman diagram I learned to decipher, consisting of two unbroken lines and a wiggle, represented a simple estimate of how two electrons repel each other. The electrical force between them is not treated as a repulsion between distant charges but as the consequence of a photon emitted by one and quickly absorbed by the other. (The effect is sometimes likened to the apparent repulsion felt by two ice skaters who vigorously throw a baseball back and forth between them; both the recoil of the throw and the impact of the catch drive them apart.) Time runs vertically upward in such a diagram, as the electrons approach, repel each other, and then fly apart. Each of the two black dots that anchor the wiggly line represents a specific point in space and time where a physical interaction takes place. More refined esti-

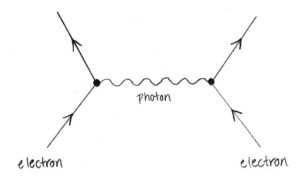

mates, represented by more complicated diagrams, resemble lacy spiderwebs of solid and wiggly lines. Each internal junction is marked by a black dot. The four dangling ends are the incoming and outgoing electrons—everything else is as firmly connected as in a real spiderweb. In the interior of the diagram, there are no loose ends.

Feynman's graphic vocabulary was eventually expanded to include other particles, such as neutrinos, quarks, and gluons, as well as the recently discovered Higgs boson. New rules and new graphical conventions were worked out. Together, the whole theory is so well confirmed in the laboratory that it goes under the imposing name *standard model of particle physics*. Machines as big as cathedrals, armies of physicists and engineers, years of work, and billions of dollars are devoted to exploring the standard model. Hitherto, it has held up brilliantly, though physicists never cease

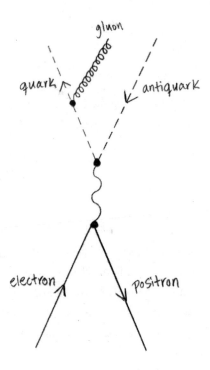

to hope that it will break down someday so they'll learn something new.

A remarkable feature common to all the diagrams depicting thousands of experiments performed in the last half century is that all external lines end on a black dot, and all internal lines have black dots at each end. That means that each individual interaction in the entire theoretical apparatus takes place at a single point in space and time—that it is strictly local. The mathematical formalism of quantum physics is explicitly local.

Locality is one of the rare properties of the world about which everyday experience, modern theoretical physics, and Einstein's intuition happen to be in perfect agreement.

So much for the mathematics. The conclusion that the underlying equations are strictly local still leaves the question of their interpretation. Einstein, Podolsky, and Rosen implied that if you insist on locality and wish to save quantum mechanics, you must give up realism.[1] QBism does that, of course, but the question remains: Where, according to QBism, are the spots, the loci in Latin, at which interactions take place? The black dots of Feynman diagrams are, after all, not actual points in space-time but mere mathematical devices used for calculating probabilities. In plain words, where, according to QBism, does stuff happen?

The QBist answer to that question is both unconventional and surprising. David Mermin, together with the original QBists Fuchs and Schack, explains: "QBist quantum mechanics is local because its entire purpose is to enable any single agent to organize her own degrees of belief about the contents of her own personal experience."[2] Personal experiences are recorded (located) in the agent's mind. They follow each other in time but by definition never occur simultaneously in widely separated locations. They are local. Their relationship to each other differs fundamentally from the connection between two masses in Newtonian gravity. A QBist cannot claim that as one body moves, a distant one feels the change because QBism only refers to a single agent's experiences.

The GHZ experiment illustrates the point. Say an agent named Alice operates one of the three widely separated detectors. From past experience she understands the correlations of the three spins, summarized in the GHZ rule. Her detector measures the vertical spin of one of a trio of electrons and finds the result *up*. Then she gets a phone call from Bob, the operator of the second detector, who reports *up* as well. If she is a conventional quantum mechanic, she can now make predictions according to either classical physics or quantum mechanics for Charlie's reading of the third detector. But not if she's a QBist. The most she can do in that case is to say: "I am very sure that when I hear from Charlie, he will report the outcome *down*." And when he does, she concludes that classical physics is wrong. She cannot "explain" the result beyond the fact that quantum theory works, but she will not attempt to tell a spooky story about it. According to Fuchs et al., "The issue of nonlocality simply does not arise" for her.

16

Belief and Certainty

On the subject of quantum mechanics, Einstein was two-thirds right. The Einstein, Podolsky, and Rosen (EPR) paper correctly suggested that the theory, as we know it today, cannot be interpreted as a local and realistic description of nature. Locality, for its part, is required by Einstein's own laws of special relativity. It was only in his insistence on some kind of physical realism that Einstein went astray.

Most people, including QBists, share Einstein's intuitive, common-sense feeling that there is a real world out there. To those who claim, on the contrary, that there are only minds and ideas, the great lexicographer Samuel Johnson offered a feisty refutation. Kicking a large rock, he exclaimed: "I refute it thus." Since his vigorous body language actually proved nothing, it is called, in analogy to an *argumentum ad absurdum* (argument to absurdity), an *argumentum ad lapidem* (argument to the stone), also known as a bald dismissal. But as an expression of a gut feeling, Dr. Johnson's dramatic gesture has a certain visceral appeal.

At issue is not so much whether reality exists but a complex of questions that have occupied scholars for ages—how we perceive that reality, how we interact with it, and how we try to represent it. Until quantum mechanics rubbed their noses in such problems, physicists managed to avoid thinking about the methods and limits of human knowledge, leaving metaphysics to metaphysicians. It is to Einstein's credit that he and his EPR colleagues at least tried to specify precisely what they meant by the word *reality*, even though their definition turned out to be too restrictive. Apparently, Einstein came to that conclusion himself, judging by the fact that after the EPR paper the phrase *element of reality* vanished from his correspondence.[1] But since the definition has the virtue of succinctness and because for a while it was good enough even for Einstein, it helps us to focus the discussion.

According to EPR, "If, without in any way disturbing a system, we can predict with certainty (i.e., with probability equal to unity) the value of a physical quantity, then there exists an element of reality corresponding to that quantity." This famous definition is couched in terms of an "if . . . then" kind of syllogism of which both the premise and the conclusion are debatable. The premise implies that a prediction that succeeds repeatedly leads to certainty. That's an example of an *argument by induction,* leading from the particular to the general. But induction doesn't have the force of a logical implication. The fact that all the swans you have ever seen are white does not prove that all swans are white. The fact that the sun has risen daily for eons is not proof that it will always do so. In fact, astronomers assure us that it won't.[2]

The conclusion of the EPR definition attempts to proceed from "certainty" to something even more substantial. If it's certain, it's supposed to be real—there is supposed to be some kind of objective physical mechanism in the real world to anchor the "physical quantity" under discussion and to guarantee the success of every prediction. But appearances—even persistent, predictable appearances—don't necessarily reveal an underlying objective truth on the ground. The everyday world, including the world of science, is too full of illusions, mirages, self-deceptions, and just plain misinterpretations to justify that conviction. Optical illusions, of which jaw-dropping examples can be found on the Internet, persuasively illustrate the gulf between facts and perceptions.

With its emphasis on amending and improving personal judgments, Bayesian probability offers an effective alternative interpretation of the meaning of certainty. A hint that "probability equal to unity" must be examined with care is inherent in the very form of Bayes' law. Recall that the acquisition of new information changes the value of a prior to a posterior value of probability by means of a multiplying factor. There is one number, though, that is never changed by multiplication—the number 0. The number 0 multiplied by any finite number is 0. If an agent assigns a prior of 0, meaning that she deems the event to be impossible or the proposition false, no amount of additional information can budge her conviction.

That the same fate befalls a prior of 1 can be proved by simply changing the proposition to its denial. Instead of asking, "What is the probability of an apple falling to the ground when it is released?" (prior probability 1), ask, "What is the probability of an apple *not* falling when it is released?" (prior probability 0), and then apply the reasoning of the previous paragraph.

Bayes' law, in short, leaves certainty unmoved. This may present a problem if the new evidence that is supposed to update a prior happens to be very strong.

Bayesian statisticians get around this defect by a simple ruse. Except for cases of mathematical or logical certainty, they simply replace priors of 0 and 1 with probabilities that are very, very close to 0 and 1 and proceed from there. The mathematician Dennis Lindley has coined the name *Cromwell's rule* for the injunction to avoid priors of 0 and 1. The reference is to a letter from Oliver Cromwell (not Thomas) to the General Assembly of the Church of Scotland imploring them not to paint themselves into a corner by justifying their convictions as immutable truths decreed by "the will and mind of God." For emphasis Cromwell uses a peculiar and unforgettable phrase: "I beseech you, in the bowels of Christ, think it possible that you may be mistaken." Cromwell's rule is an appeal to the humility, open-mindedness, and skepticism that characterize the enterprise of science, or ought to.

QBists heed Cromwell's entreaty in a very different way from Bayesian statisticians. Instead of changing the numbers themselves, they amend the interpretation of "certainty." Since wavefunctions such as *qubits* do allow probability values of 1 and 0, QBists reinterpret those values. What does it mean when an agent assigns probability 1 to an event? In the context of Bayesian probability, all it implies is that she is very, very sure that it will occur and that she would pay any amount less than a dollar for a coupon worth a dollar if the event occurs. It does not imply anything about someone else's probability assignment for the same event or about the actual makeup of the real world.

Cromwell's rule reminds me of a misconception most of my students in introductory courses seemed to share. They agreed when I said that the number 0.999 . . . , in which the three dots represent a recurring decimal, is very, very close to the number 1. But when I would go on to ask, "Do you think it is a tiny bit smaller than 1; in other words, is it mathematically correct to write 0.999 . . . < 1?" their answer was usually yes.

Not so, I would counter. A "a tiny bit" is not an acceptable mathematical term. In fact, the correct answer to my question is no, and 0.999 . . . = 1. (To convince yourself, use long division to find $1/3 = 0.333\ldots$ and then multiply both sides of the equation by 3.)

Mathematical novices are usually surprised to learn that in decimal notation the number 1, and a lot of other numbers as well, can be written in two very different ways—provided you let your mind swoop out to infinity and back again. Imagining the row of 9s without end, a process mathematicians call *going to the limit,* is a mental excursion not available to computers. An actual calculation, whether manual or electronic, truncates the infinite sequence and results in a correct inequality such as $0.999 < 1$, which does not involve a recurring decimal.

The equation $1 = 0.999\ldots$ serves as a kind of shorthand reminder of three different ways of dealing with certainty. The left-hand side is as real and concrete as your index finger—it represents the presumption of absolute certainty, which, according to EPR, is guaranteed by an element of reality. It is simple and real and finite. The right-hand side is an abstraction as elusive as the concept of infinity itself and helps to illustrate the QBist interpretation of certainty. The recurring decimal has exactly the same outward appearance as all the real numbers between 0 and 1, which are all available for representing probabilities. Symbolically, the notation $0.999\ldots$ removes the special status that

EPR conferred upon the number 1, even though the two numbers are equal. A third way to think of certainty is to omit the dots and turn the equality into the approximation $1 \approx 0.999$, which represents Cromwell's rule. Thus, the three symbols $1, 0.999\ldots,$ and $0.999,$ respectively, serve as metaphors for the way EPR, QBists, and Bayesian statisticians interpret the seemingly unproblematic notion of certainty.

According to QBism, probability 1 and 0 assignments are personal beliefs of agents, not statements about the real world. This startling conclusion brings those assignments into line with all other probabilities. There is not, contrary to the EPR definition of reality, a qualitative jump between a probability close to 1 and a probability equal to 1, no quantum leap across a boundary from uncertainty to certainty, no shifty split to overcome, and no sudden transition from opinion to fact. The degree of my belief that an apple will fall when I let go of it is numerically much greater than the degree of my belief that it will rain this afternoon, but the two judgments, though *quantitatively* light-years apart, are *qualitatively* the same.

This realization is one of the most radical consequences of QBism and possibly "the hardest principle of QBism for physicists to accept."[3] Long ago the members of the General Assembly of the Church of Scotland found it just as difficult to doubt their own judgment, which they justified in the name of their religion. They rejected Oliver Cromwell's passionate entreaty not to base certainty on belief. In our time QBism makes a stronger claim. It maintains that even certainty is a form of belief.

IV

The QBist Worldview

17

Physics and Human Experience

L ong before the invention of QBism, conventional quantum mechanics hinted that human perceptions must be hidden somewhere in its mathematical guts. The paradox of Wigner's friend shows why. If two friends don't have the same information about a quantum system, they must assign different wavefunctions to it. Since their information—what they know—is determined not only by the system itself but also by their own past experiences, those separate personal experiences directly influence their models of the world.

In 1961, near the end of his lifelong struggle to put his finger on the real meaning of quantum mechanics, Niels Bohr wrote: "Physics is to be regarded not so much as the study of something a priori given, but as the development of methods for ordering and surveying human experience."[1]

By the "a priori given," Bohr meant the external world—what Einstein called "reality." It is the rock kicked by Dr. Johnson. Notice that Bohr did not entirely eliminate the objective in favor of the subjective. What he calls the "a priori given" is not irrelevant—its role is just *not so much* at the heart of science as people have been taught to assume. While the experimenter, the observer, and the theorist are investigating *something* external to themselves, what they are dealing with directly is not nature itself but nature reflected in human experiences.

Bohr's epigram, like many of his oracular pronouncements, fell on deaf ears. I certainly never heard a word said about human experience when, as a student, I was learning quantum mechanics. Even if I had heard of Bohr's remark, I probably wouldn't have understood it. Not only because it contradicts everything I had been trained to believe about science but also because his words are obscure at best. What exactly are those "methods of ordering and surveying human experience" supposed to be? Conventional quantum mechanics provided clear and explicit recipes for systematically surveying and mapping the material world in mathematical terms, from the microcosm of elementary particles to the macrocosm of the universe—but the impressions, thoughts, and memories of the human beings who make and use the map had been carefully airbrushed out of the equations. If Bohr was right, where could those subjective elements be found in the formalism?

Forty years after Bohr's death, QBism finally managed to come up with a straightforward way to give meaning to his cryptic pronouncement. The key to the implementation of his insight is the concept of probability. According to QBism, probability—that central pillar of quantum theory—is not a thing. It is not an a priori given as frequentist probability suggests. A statement such as "The probability of a fair coin coming up heads" seems to be independent of any human influence. It stakes a claim to being a "fact." But QBism demonstrates, both logically and empirically, that probability should more effectively be regarded as a degree of belief and thereby dependent on an agent's experience. By switching from frequentist to Bayesian probability, QBism injects human thoughts and beliefs into the austere mathematical framework of physics.

QBism agrees with Bohr but takes one giant step further. Unlike Bohr it speaks not about human experiences in general but about the experience of a single agent, of a particular human. Who is that person, then? By way of emphasis, Chris Fuchs answers with the exuberant refrain from a 1970 Beatles song: "I-I-me-me-mine." He means each individual user of quantum mechanics, separately and independently. According to QBism, quantum mechanics provides a method for agents to survey and organize their own personal experiences.

If that sounds like a recipe for anarchy or a bizarre form of self-centeredness rather than a foundational principle for the grand enterprise of science, it is because we have become accustomed to misrepresenting the scope of our experiences of science. The

QBist interpretation implies a *narrowing* and at the same time a *broadening,* in a different direction, of what quantum mechanics, and by extension all of science, is about. It represents a radical narrowing because QBism restricts the relevance of a probability estimate to a single agent. But at the same time, QBism implies an immense broadening because included among the experiences of that agent are not just measurements of this electron's spin or that laser beam's frequency—"piddling events" in the greater scheme of things, to use John Bell's dismissive phrase—but all personal experiences, past and present.

Although I, as an agent, have considerable freedom to assign probability estimates to my own future experiences, they must conform to the restrictions of the calculus of probabilities. They must be free of mathematical contradictions. If, for example, I believe that there is a 20 percent chance of drawing a king in a card game, I would be foolish to simultaneously assign a probability of 30 percent to the chance of getting the king of spades. Conversely, I cannot consistently predict a 10 percent chance of drawing a face card. Psychologists and economists have demonstrated that based on faulty intuition most of us routinely flout the formal rules of probability that forbid such nonsense. In psychological experiments people have expressed absurd beliefs, such as the estimate that during a given time interval more murders occur in Detroit than in

Michigan. Such paradoxical behavior can entail dire financial and social consequences, but it seems to be part of the human condition. In science, however, it must be rooted out so the enterprise doesn't self-destruct by self-contradiction. The succinct language of mathematics helps to ensure logical consistency because its terms are much more transparent and unambiguous than those of ordinary speech.

The web of probability assignments for the totality of a particular agent's experiences differs from that of all other agents in the world. Webs of probabilities, like snowflakes, are both intricate and unique. But what about consistency among agents? If every agent lived in his or her private cocoon of personal probabilities, each internally consistent but with no agreements or consistency among them, science would dissolve into an incoherent babble of personal preferences. The broadening of the scope of what is considered to be a scientific experience provides the ties that bind science into a powerful product of human ingenuity. What connects me with my colleagues and collaborators, indeed to the rest of the scientific community past and present, is the sum of the personal experiences I have of communicating with them. Every book and article and letter I read, every lecture I hear, every conversation I participate in, every image I see, and every measurement I witness—all are new experiences added to my conscious mind, and all serve as background information for updating my probability assignments. So while each agent's collection of experiences is unique, each includes a large common core of identical shared experiences. For example, we all know and respect Newton's laws because we have all learned them and used them to calculate the prior probabilities for our future experiences. Huge patches of overlap between personal webs of probability assignments, based on shared experiences, bring order to science. Small individual differences make for innovation and progress.

Quantum mechanics, according to QBism, is not a description of the world but a technique for comprehending it. Our future experiences can only be described in terms of probabilities. They might be classical or quantum probabilities, depending on circumstances, but they are all Bayesian probabilities. An electron, for example, may be thought of as a quantum system with a spreading wave function in one experiment, but under different circumstances its motion may be likened to that of a golf ball. Conversely, Wigner would think of his friend as a classical object until, in the context of a quantum experiment, he would be compelled to construct a wavefunction that entangles his friend with an electron.

Developing an encompassing, consistent worldview is a formidable undertaking. The journey is long and hard, but QBism has shown us how to go about it.

My own adoption of QBism as the basis for a new worldview brought a deep sense of fulfillment. It put me—finally—into personal contact with the laws of nature and the people who invented them. It entangled me in the grand epic of physics in a way I had never anticipated or even thought possible.

I no longer participate in scientific research, but if I did I would continue doing what I have been doing all my life. I would handle quantum mechanics as the reliable tool that it is, compute wavefunctions, deduce probabilities from them, and urge my experimental colleagues to compare those to experimental data. But my feelings about the process have changed.

What I have been pursuing has been research according to "the scientific method." Nowadays some elementary school classrooms are adorned with a colorful wall chart that lists six or seven steps of the scientific method in more or less standard form: "1. Think of a question. 2. Do the background research. 3. Make a hypothesis. 4. Perform an experiment." Why isn't there a poster next to it titled The Artistic Method? Because even philosophers haven't been able to define art, let alone its methods, in a universally acceptable way. The enterprise of art is just too human to be described on a poster. It involves feelings and idiosyncrasies and individual differences in such an essential way that forcing the "artistic method" into the straitjacket of a definition is not only impossible but also counterproductive. If it could be designed, an artistic method poster would surely inhibit rather than inspire the children who memorize it.

If the artistic method is too human for standardization, the canned scientific method suffers from the opposite problem. There is no room for individuality or personal differences in its standard description. It sounds more like a set of instructions for operating a lawn mower than a portrayal of a glorious human adventure.

QBism offers a more appealing point of view. By placing the users of quantum mechanics, each agent individually and personally, at the center of the action, "QBism puts the scientist back into science," as David Mermin wrote in 2014 in the journal *Nature*.[2] For me this implies that as a physicist I am not merely following a set of rules that have evolved for millennia without my participation. Instead, QBism allows me to feel that I am working independently, guided by my own experiences and thoughts—which, of course, have been informed and nurtured by those of my illustrious predecessors. In the end what matters are my own personal probability assignments. QBism has internalized them and thereby humanized science.

QBism implies a radical change of perspective. It turns the traditional top-down view upside down by offering a bottom-up depiction of the universe.

Conventional physics, which strictly separates object from subject, attempts to see the world from a universal vantage point. The laws of nature are fixed and immutable. The material universe exists "out there," ruled by those laws and unaffected by us puny humans who contemplate it. Time too is objective, in the sense that it is divorced from personal feelings, beliefs, and points of view, even though the effects of speed and gravity described by the theory of relativity complicate its flow. Human understanding, in this way of thinking, while never reaching God's, nevertheless aspires to capture bits of divine wisdom. In the next four chapters, I will explore how QBism revises this worldview and replaces it with a more humble one, which, instead of arguing from the general to the specific, seeks to find the universal in particular personal experiences.

18

Nature's Laws

> This is one of man's oldest riddles: How can the
> independence of human volition be harmonized
> with the fact that we are integral parts of a uni-
> verse which is subject to the rigid order of na-
> ture's laws?
>
> —Max Planck, *Where Is Science Going?*

The laws of nature are not revealed simply by looking for them the way we discover a new planet or a new species of ant. Instead, they are freely invented on the basis of a limited number of observations and experiments. Their formulation, as Planck knew from hard-won experience, requires not only logic and mathematics but imagination, intuition, insight, and instinct. The method of finding those laws uses induction—reasoning from the specific to the general—a procedure that is as fallible as any human endeavor.

Unlike a simple observation, which can be recorded and shared with others as soon as it is made, a new principle of science starts as a hypothesis (or a guess, as Richard Feynman called it in his blunt way) and then requires an extended period of trial before it achieves the lofty status of a "law of nature." Take Newton's formula for gravity, for example, one of the first postulates of physics to earn the designation *law.* Initially distrusted and occasionally even mocked, gravity took decades to be accepted by science and the public. Corroborated by one successful explanation after another (ocean tides, Earth's equatorial bulge, predictions of eclipses and comets . . .), it slowly gained credence, climbing toward certainty and popular acceptance.

The way hypotheses solidify and crystallize into laws resembles the transition from predictions to elements of reality in the reasoning of Einstein, Podolsky, and Rosen. In both cases a belief gains strength and gradually leads to certainty. Once a principle is endowed with the prestigious label *law of nature,* its meaning begins to change The law starts to be called upon not only to describe the way things happen but to control or govern them. It begins to rule the world in the sense of the phrase "the rule of law." Or conversely, as Planck put it, the universe becomes subject to the rigid order the law imposes.

We know where human laws come from and how they are made, but where do nature's laws come from? For believers like Newton, God decrees the laws and to the extent that we can figure them out, we learn to understand and appreciate a smidgen of the mind of God. On this view nature's laws are divine laws, and that's all there is to it. Unfortunately, religious explanations tend to shut off debate rather than stimulate curiosity and discovery.

For classical physicists like Planck, as well as the majority of my colleagues today, nature's laws smell of the absolute. To be sure, we all know and accept the fact that scientific theories evolve and mutate and are subject to possible recall, but until proven wrong, laws are assumed to hold absolute sway. The laws of special relativity, for example, are absolute, as contradictory as that may sound. They have never been found to be violated and are universally accepted as valid. Unless and until they are convincingly found to be in error, all physical theories must be brought into compliance with special relativity. In the same way, all of nature's laws are absolutely valid—until further notice.

The notion that nature's laws control the world pervades science teaching. When a schoolchild is asked why a puck continues to slide on the ice instead of stopping as soon as it loses contact with the stick, she is supposed to answer along the lines of: "Because of the law of conservation of momentum." The law is believed to command inanimate matter; the puck merely does what it is ordered to do by an all-powerful master—a law of nature. In that sense the law "causes" the puck to remain in motion—just as traffic laws cause drivers to obey speed limits. But since a puck doesn't have free will, the natural law it "obeys" must differ in some profound way from a highway speed limit.

So what is the status of a law of nature? Where does it come from? Who wrote it? Where does it reside? Is it somehow encoded in matter or in the space-time of the universe? How is it enforced? Did it operate before it was stated for the first time? If we don't know where a law came from, isn't it really a miracle—the way Newton's law of gravity was a miracle? Are the laws of nature themselves supernatural, above and beyond the reach of science?

Often, in science as in life, you learn a lot about what a thing is by studying how it came to be; the history of a phenomenon reveals clues to its meaning. Since laws of nature are born in the minds of scientists, it may be that we should look there for clues to their essence, rather than in nature itself or beyond it on some higher plane.

QBism's answer to the question of the status of nature's laws is more down to earth than any religious or supernatural explanation. The Bayesian interpretation of probability as a measure of expectation of future experience suggests that the tradition that has elevated the laws of nature to their current transcendent status has it backward. QBism implies that things don't happen the way they do because they obey a law of nature but that the laws of nature have been invented because things happen the way they do.

Laws of nature thereby take on a new role. Rather than determining events, they describe the past experience of events. They are supremely efficient summaries of information; shining examples of what computer scientists call *data compression.* The amount of scientific information contained in the eight little symbols that comprise Newton's law of gravity is unimaginable in its scope—as unimaginable as the infinite sequence of digits that define the number that is succinctly described as "the square root of two." Regarded as a summary of information, the word *law* seems inappropriate. Perhaps the word *rule* comes closer to expressing its meaning. (The word *rule* comes from *regula,* meaning a straight stick.) A rule can be interpreted as an observed regularity rather than an edict imposed from above—even though it may be as fundamental and inflexible as a law. As part of the laws of electromagnetism, for example, the so-called right-hand rule describes the direction of the magnetic field that surrounds a current-carrying wire. That rule is as rigid as a traffic law, but it has a humbler name.

In the QBist worldview, nature's laws gain credence asymptotically, approaching certainty ever more closely at a diminishing rate of change. Just as the probability that a radioactive atom has decayed rises from 0 to 1 but never gets there (unless it is observed), the probability that a law of nature is in fact valid rises from 0 (before the first guess was made) to 1—without ever reaching certainty. Cromwell's rule should apply not only to probabilities but to nature's laws as well. By allowing an infinitesimal sliver of doubt to temper their absolute validity, we become better prepared for the refinements and qualifications that are bound to update them in the future.

I'm very, very sure that the cup resting on the table in front of me will not spontaneously levitate toward the ceiling—but I believe it would be imprudent to claim that I'm absolutely sure. I would bet money on my conviction, but I would insist on reserving a sliver of doubt. In fact, even classical physicists conceive of a minuscule chance that an accidental and extremely rare accumulation of air molecules under the cup could lift it like a balloon.

QBism has taught me to regard the laws of nature that I have been teaching for half a century in a new light. These laws represent the experiences and the wisdom compiled by generations of physicists, but they are neither absolute nor rigid. They are human creations and therefore malleable—at least in principle.

The QBist interpretation of the nature of nature's laws frees us from the iron grip of rigid determinism that Planck was alluding to in this chapter's epigraph. But what does QBism say about the antithesis of strict determinism—about the possibility of human volition and free will?

19

The Rock Kicks Back

The American theoretical physicist John Archibald Wheeler (1911–2008) is better known to the general public for enriching our language with the term *black hole* than for his pioneering contributions to nuclear physics. In the scientific community, he was regarded not only as a bold and imaginative theorist but also as an inspiring teacher. His most famous student was the enfant terrible of American physics, Nobel laureate Richard Feynman, whose PhD dissertation he supervised. Forty years later at the University of Texas, Wheeler served as undergraduate research adviser to Chris Fuchs, whom he encouraged to pursue the study of the foundations of physics, which at that time most of us regarded as a fringe topic at best. From his teacher Fuchs learned that quantum information may turn out to be the most promising key to a deeper understanding of quantum mechanics itself and, by extension, of physics in general. Accordingly, John Wheeler may be called the godfather of QBism.

Wheeler liked to pose what came to be called Really Big Questions (RBQs) in cryptic, oracular language. Among them were, Why the quantum?, It from bit?, and A participatory universe?

The first question is as elusive today as it was for Max Planck. At the beginning of this book, I proposed that Planck's $e = hf$ was the icon of quantum mechanics. Where does it come from? At the time it was an unsupported hypothesis, whereas today it follows from the more fundamental, and more complicated, principles of quantum mechanics. But what is the simple essence of those principles? Perhaps this RBQ is truly profound, or maybe it has no answer, or most likely, it isn't phrased properly. If, for example, the world is really quantum mechanical at its ineffable core and we just don't happen to notice that in our classical everyday world, then the question might be turned around. If the quantum is as unexplainable as existence itself, the real question may be, why the classical? In any case by asking *why* rather than the more pusillanimous *how,* Wheeler signaled his penchant for metaphysics. Philosophical questions about the meaning of being and reality should, he felt, regain their rightful place within the discipline of physics, from which they had been banished for centuries. Chris Fuchs has taken that advice most emphatically to heart.

The second RBQ, It from bit?, which Wheeler in his more assertive moods used without the question mark, is an extreme example of data compression. In three tiny words, it summarizes the entirety of Wheeler's philosophical legacy, which proposes information as the key to understanding nature. Is the *bit,* regarded as an atom of information, even more fundamental than the chemical atom for our understanding the *it*—the material universe? QBism is this century's first, and surely not last, chapter in the grand metaphysical quest called It from bit.

With his most radical question, A participatory universe?, Wheeler underlined the lesson we learn from quantum mechanics—that experimentation and measurement are not the acts of a passive, detached observer examining an external, independently existing world, as classical physics had assumed since the time of Democritus. Instead, the observer is intimately involved with the object under study. Rather than acting as mere recorders of information, we are agents who participate in the very creation of the outcomes of our interactions with the world.

QBism answers Wheeler's question in the affirmative and elaborates. From its very beginning, quantum mechanics has been preoccupied with physical experiments called *measurements.* Typically, an apparatus is set up to measure some attribute of a quantum system, such as the spin direction of an electron. Then, a wavefunction is calculated for the purpose of predicting the probabilities of outcomes, the experiment is performed, and the empirical data are compared with the predictions.

Many physicists have objected to the word *measurement* by pointing out its misleading connotations. The word seems to imply that the value of the outcome preexists the experiment and is just waiting to be revealed. Measuring the weight of a baby, for example, tacitly implies that the baby has a weight, which just happens to be unknown. The measurement merely pulls off the veil over that value and lays it bare for all to see.

In quantum mechanics, however, *unperformed experiments have no outcomes*. The electron spin has no direction until we determine it. The *qubit* representing spin has no bit value until the wave function collapses to *up* or *down*. In fact, if we assume that there is a hidden spin value, we are led into error—as the GHZ experiment so dramatically demonstrates. It's not a matter of not knowing which spin value to assume. What's wrong is the fundamental assumption that a spin value exists at all.

According to QBism, a measurement does not reveal a preexisting value. Instead, that value is created in the interaction between the quantum system and the agent.

Chris Fuchs explains:

> QBism says when an agent reaches out and touches a quantum system—when he performs a

quantum measurement—that process gives rise to birth in a nearly literal sense. With the action of the agent upon the system, something new comes into the world that wasn't there previously: It is the "outcome," the unpredictable consequence for the very agent who took the action. John Archibald Wheeler said it this way, and we follow suit: "Each elementary quantum phenomenon is an elementary act of 'fact creation.' "[1]

That's what Wheeler meant by the term *participatory universe.* As we live and go about our business, we not only interact with the universe—we continually participate in its creation.

That sounds arrogant, but QBists don't really claim credit for creating the universe. Quantum mechanical experiments create only minuscule, practically invisible additions to the fabric of the world, not the whole shebang. They serve the important function of demonstrating what is possible. Furthermore, even the smallest causes can have momentous effects, as chaos theory has convincingly demonstrated. (Key phrase *butterfly effect:* the flap of a butterfly's wing in Mexico may ultimately cause a hurricane in Texas.) But while allowing for such potential leveraging effects, the bulk of the universe obviously came into being without the help of experimental physicists. Just how it did remains to be worked out, but John Wheeler, ever the visionary, doesn't hesitate to go for broke:

> It is difficult to escape asking a challenging question. Is the entirety of existence, rather than being built on particles or fields of force or multidimensional geometry, built upon billions upon billions of elementary quantum phenomena, those elementary acts of "observer-participancy," those most ethereal of all the entities that have been forced upon us by the progress of science?[2]

The phrase "elementary act of observer-participancy" is misleading. Participating in quantum experiments and observing their outcomes is how physicists stumbled upon quantum theory, but the elementary acts involved may be much more common than measurements in physics labs. If an observer—an agent, along with her apparatus—is regarded as a large quantum system, then an experiment is in essence an interaction between two quantum systems, and we have learned that it results in the creation of new facts. The same kind of fact creation occurs when any two quantum systems happen to come together. That, according to Wheeler, may be the mechanism for the evolving creation of the universe: quantum systems collide and interact and thereby create new "facts on the ground." Wheeler left his third RBQ for future generations to answer. Chris Fuchs and the QBists have taken the first step toward an answer.

Defenders of the QBist interpretation of quantum mechanics are often criticized for their take on "reality." Because they consider wavefunctions and the probabilities they yield to be unreal, QBists are accused of denying reality altogether—an unfounded, illogical charge.

In fact, QBists firmly believe that the real world exists out there, external to ourselves. But instead of insisting that scientists are mere detached observers and recorders of that reality, they count themselves as part of it and active participants in its formation. In an act of observer participancy, there is no dominant party—the observer and the observed participate on equal terms. Potentially, therefore, every particle, along with every agent, is a participant in the creation of the universe.

Understood in this way, the QBist universe is not static but dynamic; less like an intricate clockwork and more like the interior of an evolving star that is not alive in the conventional sense but bubbling with creative energy and continual surprise. It is real but veiled, objective but unpredictable, and substantial but unfinished.

Far from denying reality, QBists believe that the evidence for what is real comes primarily from quantum mechanics itself. Fuchs expresses it this way: "We believe in a world external to ourselves precisely because we find ourselves getting unpredictable kicks (from the world) all the time." By way of illustration, he mentions a typical experiment. An agent sets up her equipment to prepare a quantum system in some special configuration determined by her own free will. She calculates subjective probabilities for the occurrence of various possibilities for the outcome of her measurement, but she can do no more than that. The external world, interacting with the apparatus, determines what actually happens in the end—which possibility is actually realized. "Thus," Fuchs concludes, "I would say in such a quantum measurement we touch the reality of the world in the most essential of ways."[3]

Dr. Johnson thought he was illuminating the nature of physical reality because he and his audience were able to predict with certainty what he would experience upon kicking his rock. The rock was an example of an Einsteinian "element of reality." But if he had kicked a quantum rock instead, he would have had to make do with predicting one of several different possible outcomes, each with its own probability. The choice of what actually happens is made not alone by the good doctor using his free will nor by the rock obeying strict laws of nature but by both of them in the act of colliding. According to Fuchs, our inability to predict with certainty what will happen in a quantum experiment thus reveals more about the real nature of the world than classical physics, with its laws and its certainties, has been able to discover. In this real quantum mechanical world of ours, the observer participates and the rock kicks back!

20

The Problem of the Now

When I was eleven years old, I made time stand still. Although I was puzzled and vaguely alarmed by the inexorable flow of time, I was resigned to the fact that I couldn't make it stop altogether. But, I wondered, could I at least stop *one* moment, a sort of reference point that would forever remain fixed? Without being able to articulate it, I probably knew from experience that the further back I cast my mind, the less clearly I could remember specific moments— so even if I selected a fixed point in the past, it might quickly blur and fade away. To make sure I would not forget it, I determined to pick a unique point sometime in the future rather than the past and fix it like a butterfly specimen on a pin.

I knew beforehand that I had to prepare the moment thoroughly by being uncommonly observant and fully aware of its context. A train trip I took alone, traveling from my grandparents' house in Basel, Switzerland, back to our home near Zurich, furnished the occasion. I knew that soon after leaving Basel the train would pass a pretty little castle in a clearing in the forest outside the right-hand window. I loved catching a glimpse of that fairy-tale château with its yellow brick walls and its crenellations outlined in ochre and never tired of the sight.

On the day in question, I prepared myself by examining and memorizing my surroundings in as much detail as I could. Today, almost seventy years later, the scene is still clearly etched into my memory. Cell phones were still far in the future, and I didn't have a camera, but my recollection is vivid. It was late afternoon on a warm autumn day, the carriage was almost empty, the battered wood seats were hard and uncomfortable, and the reassuring clickety-clack of the wheels, so familiar from countless outings, would have lulled me to sleep if I hadn't been so focused on the task at hand. As we emerged into the clearing and the castle appeared quite close to the tracks, I concentrated as hard as my boyish attention span permitted and, in the instant we passed, shouted, "Now!" If my fellow passengers were startled, I didn't notice. I had captured my moment. I had stopped time.

Years later I was mildly disappointed to learn that the building is really part of the Feldschlösschen brewery and shows up on the labels of its beer.

Of all the subsequent memorable moments in my life, none have had the distinction of being intentionally selected for the purpose of stopping time in its tracks. Occasionally, in a lecture about time, I would tell the story of the Feldschlösschen and lead the audience in a recreation of the experiment. We would preview what we were going to do, and why, all the while building up anticipation for the coming moment until we could almost physically feel its approach— closer and closer as if it were moving toward us in some ineffable kind of motion. Finally, we would count down together from ten to one, finishing by shouting "Now!" in unison. Afterward we would refer back to that moment and try to describe the similarities and differences between the actual instant and its rapidly receding memory. My students loved the exercise, but for me those later Nows lacked the intentionality and novelty of the original. In my mind the later recreations have lost their focus and faded away. Once, to my delight, I ran across a student who remembered the experience ten years after the fact.

The trouble with time is that it doesn't exist. The past is gone and leaves only traces in memories and records. The future hasn't come into being yet. If a particle's journey through space and time is depicted by a meandering line in space-time, the meeting point between past and future—the present—is only that: a point on the line. Just as a point in space lacks extension, a point in time lacks duration. It's a mathematical idealization, an abstraction, an idea.

And yet, the present is the only direct experience we have of time. When we think of the past, we are conscious of delving into a memory bank. When we think of the future, we realize that we are anticipating something yet to come. But the present is here with us. We feel it in our bones and every one of us knows exactly what it is because we're in it. In fact, according to Buddhist teaching as well as modern psychological lore, being fully in the present is a recipe for mental and spiritual well-being. The enormous psychological significance of the Now makes its representation by a point seem almost ludicrously inadequate.

Albert Einstein, who had gone to school in Aarau, a scant twenty miles across the hills behind the Feldschlösschen, had worried about the Now. In part that was because he had made the problem worse. By rejecting Newton's absolute time, which is the same everywhere in the universe, and substituting relative time, which depends on an observer's motion and gravitational environment, he had sown confusion into the very definition of the meaning of the Now. But his concern was more fundamental. The philosopher Rudolf Carnap recalled a conversation in which Einstein explained that "the experience of the Now means something special for man, something essentially different from the past and the future, but that this important difference does not and cannot occur within physics. That this experience cannot be grasped by science seemed to him a matter of painful but inevitable resignation."[1]

Cornell physicist David Mermin realized that even though the distinction between past, present, and future is somewhat peripheral to the interpretation of quantum mechanics, QBism offers a convincing resolution to the problem of the Now. When the content of physics is understood in terms of personal probability estimates—with values that may reach 0 and 1—for future experiences of agents, the Now, like every other human experience, is unique to each agent. If I depict myself as a point in space-time, I can draw a meandering line representing my position and the reading on my watch as I go about my business. The time on my watch at which memories and records of my environment are established is the advancing moment called Now. The line, with its point inexorably advancing as time elapses, is divided into two portions called my past and my future, and they meet at my Now. That diagram and its interpretation are perfectly acceptable to physics.

QBism brings two new insights to the story. It speaks explicitly of human experiences, which classical physicists like Einstein considered to be off-limits to science, and it reminds us that the map is not the territory. Although I, as agent, represent my body as a point, I know very well that I'm not a point. Nor is my Now, which I experience vividly as I type this but which, by the time you read this, has long since vanished, a point. My body and my Now are no more points than an electron is a *qubit*.

Mermin goes on to explain that the Now experience, even though it is private, can be shared, just as Wigner and his friend shared the experience of measuring an electron spin. Because physics is local, the sharing agents must be in close proximity—close enough that the time it takes for a signal to pass between them is negligibly small. In that case the complications of relativity theory do not even arise. Two observers or agents at the same location agree that their Nows coincide as long as they are together. So when my wife and I share a glass of wine in the evening, we experience the same Now together. And when I shouted, "Now!" with my classes, we were all, for a moment, in the same Now. But when my students wandered off to distant points, our Now diverged into separate experiences with nothing in common.

As usual Einstein was ahead of his time. In expressing his frustration with the inability of physics to deal with the Now, he was putting his finger on a topic of profound interest to future generations. Mermin's take on the meaning of the Now is straightforward and convincing, but it represents only the simplest possible approximation of a subtle and rich phenomenon. Just as the period at the end of this sentence turns out to be a teeming world of interacting particles when seen under sufficient magnification, my Now, under close examination, is a wonderfully complex and meaningful phenomenon. Before physics can contribute anything to our understanding of it, neuroscience, which deals with cells and electrical currents at the classical rather than the quantum level, will have to weigh in.

My Now, whatever it is, is an experience that the context in which it is embedded determines in large part. Much of it has to do with my immediate physical surroundings, my *here,* such as the railway carriage and the view out of its window in the experiment of my youth. Some of that stage set entered my consciousness directly through my eyes, but much of it was stored in the immediate memories of what was behind, above, and below me, out of sight. Beyond the visual setting were the sounds, smells, and emotions I experienced just moments before that particular Now.

Most intriguing, though, is the realization that the anticipation of experiences in the immediate future influence the Now. The fuzziness surrounding the point in time we call the present extends not only backward, through memory, but a little ways forward too. The brain is not, as is commonly believed, merely a reactive organ. It is also in large part a predictive organ. Constantly, without our awareness, our minds are making an astronomical number of predictions of what is likely to happen next. Even an action as simple as reaching for a cup of coffee involves the fast and intelligent (nonrandom) control of fifty or so different muscles in my hand and arm—a computational task of staggering complexity, without which the hand would miss the cup. That silent buzz of activity enables us to function in the world.[2] It is a hidden part of the Now.

Bayesian methods provide the most natural way to describe human motor control. Past experience furnishes the prior probabilities for predicting how the body's cells are going to react to specific electrical impulses and then actual sensory inputs (measurements, in physics terms) update the priors via Bayes' law. The updated probabilities in turn guide the subsequent nerve impulses that control the muscles.

If such a model of human perception turns out to be successful, a microscopic examination of the Now will be found to fit neatly into the QBist worldview. Evolution will be found to have anticipated science. And eventually, Einstein's dream of including the present moment into the framework of physical science may be fulfilled in ways he could not have imagined.

21

A Perfect Map?

I n one of his last novels, Lewis Carroll, the author of *Alice in Wonderland,* described an ideal map:

> And then came the grandest idea of all! We actually made a map of the country, on the scale of a mile to the mile! But its sheer size caused problems: It has never been spread out, yet . . . the farmers objected: they said it would cover the whole country, and shut out the sunlight! So we now use the country itself, as its own map, and I assure you it does nearly as well.[1]

Physicists are more sophisticated. Since the days of Newton, the idea of a perfect mathematical model, analogous to a perfect map, has been the ultimate goal of physical science. With full awareness that the map is not the territory and taking advantage of the remarkable ability of mathematics to compress data, the physicists' perfect map should be one-to-one, not quite like Lewis Carroll's but in the following sense: every feature of the physical world should have a counterpart on the map, with nothing left out, and every element of the map should in turn represent a part of the real world. For example, the atomic hypothesis that matter consists of atoms and the void was a patch of that perfect map and so was Newton's law of gravity.

A perfect map would depict God's view of the world. If we humans understood it, we would know the mind of God. A perfect map is a distant goal that even classical physics is incapable of achieving. It is not only impossible to record the position of a particle with infinite precision, but our study of chaotic systems, which the development of the computer spurred on in the last quarter of the twentieth century, demonstrates an even more troubling problem. In most physical systems, it turns out that even if we fixed the coordinates with a very small error, the discrepancy between our mathematical prediction and the actual configuration of the system would rapidly grow to unacceptable levels. In other words, prediction far into the future is not possible in realistic systems.

In classical physics a perfect map is impossible to achieve as a practical matter, but it is still conceivable as a theoretical ideal. Even if *we* can't, God may see the world that way from on high, and we can strive to approach his point of view. But QBism, with its intrinsic randomness and its Bayesian probabilities, puts an end to the hope that we can ever know the mind of God.

Quantum mechanics, to the extent that it has been experimentally corroborated, forces us to admit that absolute certainty in prediction cannot be achieved, and QBism, to the extent that it furnishes a reasonable interpretation of quantum mechanics, implies that science is not about ultimate reality but about what we can reasonably expect. For many, including Einstein, giving up the quest for a perfect map meant a melancholy admission of defeat, but Marcus Appleby, whom we met in Chapter 9, has a much more optimistic view of the matter.[2]

First off, he points out that QBism doesn't detract in any way from the immense success of quantum mechanics in helping us understand not only the material world but, through biochemistry and neuroscience, the foundations of the life sciences as well. Knowing what we can reasonably expect and how firmly we should expect it is as close as we can come to understanding and controlling the world.

Appleby's second point is that QBism, by moving physics closer to human thoughts and feelings, may have a better chance than raw materialism to solve the ancient enigma of consciousness, the problem of the relationship between the mind and the brain. He admits that at present that's only a hope. Appleby's conclusion, though, is as surprising as it is delightful:

> The ambition to "know the mind of God" is not realistic. But I would go further than that. I would question whether the idea is even attractive. Suppose one really could comprehend the universe

in its entirety. Might this not be found a little cramping? If the universe really could be comprehended in its entirety it would mean that the universe was as limited as we are. It seems to me that living in such a universe would be rather like trying to swim in water that is only six inches deep. . . . My personal feeling is that I would not wish to belong to a universe that I was able to fully comprehend. Against this vision, of physics as knowing the mind of God, I would like to set another: physics as swimming in water that is a great deal deeper than we are—perhaps even infinitely deep.[3]

If, in contrast to Appleby, we persist in lamenting our inability to find a perfect map, we can take comfort from Lewis Carroll's advice: For guidance in finding our way around, the territory itself serves nearly as well. QBism reveals how. Our experiences of the territory—the external world—furnish the clues we need for figuring out what we can reasonably expect to find around the next corner. Who needs more?

22

The Road Ahead

I n his 1965 Nobel Prize lecture, Richard Feynman related the real journey—blind alleys, detours, wrong turns, and all—of his contribution to the development of quantum electrodynamics (QED), the fundamental theory of electrons and photons.[1] In the course of this quest, he learned to appreciate the value of expressing a theory using different mathematical formulations, which in the end turn out to be logically equivalent. He knew, for example, that quantum mechanics can be couched in the language of wavefunctions or matrices, and he devised yet a third scheme based on ensembles of classical trajectories, which on the surface resembles neither of the first two. Even the venerable nineteenth-century classical theory of electricity and magnetism received a radical makeover by Feynman.

The point of stating the same thing in different terms is to deepen understanding. In my teaching career, I have learned the painful futility of "explanations" of difficult topics that amount to repeating the same words over and over again. Fresh turns of phrase and novel mathematical frameworks expressing the same essential meaning inevitably bring with them new allusions, images, and overtones, which in turn enhance comprehension. Thus, when Feynman set out on the monumental task of combining electrodynamics with quantum mechanics, his mathematical tool kit contained not only standard versions of the two theories but several equivalent variants of each.

Feynman, being Feynman, dug deeper. What is the meaning of these multiple reformulations? "It always seems odd to me," he said, "that the fundamental laws of physics, when discovered, can appear in so many different forms that are not apparently identical at first, but, with a little mathematical fiddling you can show the relationship . . . I don't know why this is—it remains a mystery, but it was something I learned from experience."

And of course Feynman suggested an answer in his Nobel lecture: "I don't know what it means, that nature chooses these curious forms, but maybe that is a way of defining simplicity. Perhaps a thing is simple if you can describe it fully in several different ways without immediately knowing that you are describing the same thing."

In that light what is this simple thing that animates the fancy formalisms of quantum mechanics? "Why the quantum?" as John Wheeler put it. QBism doesn't answer that question—yet. Along with others listed in the Appendix, QBism is a new *interpretation* of the existing theory, not a *reformulation* in the sense of Feynman. QBism is important and powerful and entails philosophical consequences of lasting significance, but it does not affect the actual technical content of quantum mechanics, which allows comparison of the theory with experiments. Only the meaning of the concepts that enter into it—especially of probability—is changed by QBism. What's lacking so far is a completely new version of the old theory.

But it's early days yet. One of the most important attributes of a new scientific idea is that it should be *heuristic,* leading to further research, inspiring fresh ideas and questions. The word *heuristic* comes from the Greek for *finding:* a heuristic idea spurs you on to new discoveries. In the very title of the famous 1905 paper in which Einstein introduced photons with energy $e = hf$, he described his proposal as heuristic.[2] The history of twentieth-century physics proved just how remarkably prescient this description of the quantum hypothesis turned out to be. QBism too shows promise of playing a heuristic role in the search for the real meaning of quantum mechanics.

QBism suggests the question: Why the wavefunction? Do we really need that abstract mathematical device, which seems to incite so much controversy and in the end must collapse before it furnishes probabilities? Couldn't quantum mechanics be couched directly in terms of probabilities, that is, real numbers between 0 and 1, bypassing wavefunctions with their nebulous status and imaginary components? If that were possible, those weird maps called wavefunctions could be discarded and banished to the attic of the history of science.

In fact, it is possible. There is no proof that wavefunctions are the *only* way to capture the phenomenon of superposition, even though, because they were inspired by intuitively accessible classical waves, they clearly do it well. At issue is not the possibility of rewriting the theory in different terms but a question of simplicity. No fundamental principle stands in the way of translating the mathematical formalism of wavefunctions into the language of probabilities—but unless it is done cleverly, the result might be a monstrously complicated, ugly-looking theory. If that were to turn out to be the case, physics would not have gained much of value. It would be a little like describing the solar system not in terms of elegant, abstract Keplerian ellipses but as a clumsy compilation of raw planetary coordinates observed directly. It would be a step backward.

Undaunted, QBists are pursuing a program of expressing the quantum rules in terms of probabilities rather than wavefunctions. In the course of this mathematical exercise, they came across an elegant and versatile way of dissecting any experimentally testable probability into a sum of basic, more primitive "standard" probabilities. (The process is reminiscent of Euclid's *fundamental theorem of arithmetic,* which allows every integer to be dissected into a unique product of primes, a procedure that has played a major role in the history of mathematics.) Recently, such standard quantum measurements have actually been performed in the laboratory and have shown to be as simple and useful as QBists have claimed.[3]

The visual appearance of the formula that relates an actual quantum probability to the standard ones came as a surprise. It looks almost exactly like the equation for a basic principle of conventional classical probability theory that furnishes the total probability of an outcome that can be realized via several distinct routes. In the case of a coin, for example, the probability of throwing heads plus the probability of throwing tails must equal 1 ($1/2 + 1/2 = 1$, for a fair coin), reflecting the fact that if there are only two possibilities, one or the other outcome is sure to occur. This is a simple case of a theorem in classical probability theory called the *law of total probability.* We tacitly used it in the Bayesian calculation of the odds of having cancer when we expressed the total probability $p(+)$ of obtaining a positive test as the sum of the probabilities for true and false positives.

In quantum theory this law does not hold in its classical form. In Feynman's beautiful experiment, for example, it is not true that the probability of finding the electron at a certain spot when both slits are open equals the sum of the probabilities for the cases in which one slit or the other is blocked.[4] Quantum probabilities don't add up—they can interfere and even cancel out. This point is so fundamental that it prompted Feynman to choose the double-slit experiment as an illustration of "the only" mystery of quantum mechanics.

It was a relief, therefore, rather than merely a surprise, that the QBists' newly derived equation, called the *quantum law of total probability,* deviates from its classical counterpart. But the two equations look tantalizingly similar, differing only in one tiny extra term—a term that is quintessentially quantum mechanical in origin. And the modification is not, as one might have guessed, related to Planck's ubiquitous constant h. In a sense the extra term is even more fundamental than Planck's constant.

Before I reveal what that little quantum deviation actually turned out to be, I must confess to a sin of omission. In science, as in life, unexpected obstacles crop up. The equation I have been describing has been proven except for a pesky, purely mathematical detail that's holding up progress. That technical gap has attracted the attention of a small international coterie of mathematicians and mathematical physicists, but though the solution is easy to conjecture, its proof is turning out to be elusive. By now it has defied a decade of effort, but in the process it has revealed beautiful hitherto unsuspected connections in pure mathematics. The mathematician Jon Yard has even suggested that the conjecture may be related to one of the famous twenty-three unsolved problems that David Hilbert posed in 1900. (That celebrated list, which has been successfully pared down to about half its original length in the intervening years, continues to challenge and inspire mathematicians.) If the conjecture is confirmed and helps to solve a Hilbert problem, or even part of one, QBism will once more demonstrate its heuristic power and will thereby command added respect from both mathematicians and physicists.[5]

Back to total probability. The term that distinguishes the quantum version of the law from its classical cousin turns out to be an integer called the *quantum dimension* of the system under discussion and denoted by the letter d. The quantum dimension has nothing to do with space or time but with the number of states a quantum system can occupy. It is the dimensionality of the abstract space in which the wavefunction operates, and it measures the size of the spreadsheet when the wavefunction is expressed as a matrix. The quantum dimension of a *qubit*, for example, is 2, reflecting the fact that the *qubit* ball has a two-dimensional surface. For the GHZ three-electron system, $d = 8$, while for other systems d can range all the way up to infinity.

The quantum dimension is an intrinsic, irreducible attribute that characterizes the quantum nature of any system. Even more fundamentally than Planck's constant it signals a departure from classical behavior. Chris Fuchs compares its physical significance to that of mass, which characterizes the inertial and gravitational properties of material objects. Quantum dimension is implicit in every quantum mechanical calculation but rarely shows up as explicitly as it does in the quantum law of total probability. It is a natural property of the material world that defies human perception, somewhat in the way that the ubiquitous relativistic distortion of space induced by mass completely eludes our senses.

If the missing mathematical proof is found, QBists will have in hand a powerful new tool. Fleshing out the practical meaning of the quantum dimension would be a big step toward answering the question, Why the quantum? At the same time, the quantum law of total probability may turn out to be the basis of a radically new formulation of quantum mechanics without wavefunctions, which will, as Feynman has pointed out, deepen our understanding. That is certainly Chris's hope. Specifically, he would like be able to install the quantum law of total probability as the principal axiom of quantum theory.

Much more speculative than this technical development is the suggestion by Marcus Appleby that

QBism, by its very nature, may provide the bridge between psychology and physics that is needed to disentangle the ancient complex of problems around human self-awareness, free will, and the mind/body relationship. At the very least, QBism frees physics from the spell of Democritus—the misleading assumption that we can truly understand the world in purely objective terms as existing outside ourselves, detached from our thoughts and feelings and perceptions. Without that first liberating step, Democritus's curse ("Wretched mind . . . Your victory is your own fall.") will continue to haunt us.

But we must be patient. Remember that from the Greek conception of atoms until today's scanning tunneling micrographs of real atoms, more than two millennia elapsed.

Appleby seems to be more optimistic about the arrival of the next stage in the history of science. In a conversation about the quest to fashion a union of physics and psychology into some kind of psychophysical synthesis, he told me that such a project might take "hundreds of years" to achieve. In the context of today's breathless pace of science, that estimate sounds like an admission of defeat, but Marcus is a mathematician and used to waiting. Fermat's last theorem, for example, was proved in 1994 after 357 frustrating years of failed attempts. Taking the long view of the promise of QBism, as Appleby does, comes more naturally to a mathematician or a philosopher than to a physicist.

Rüdiger Schack, a member of the original QBist triumvirate, is more confident. "Let me finish with a prediction," he said in an interview in 2014. "In twenty-five years when a new generation of scientists have been exposed to QBist ideas, QBism will be taken for granted and quantum foundations will have disappeared as a problem."[6]

What to do in the meantime? Max Planck, whom Schack was echoing, famously remarked: "A new scientific truth does not triumph by convincing its opponents and making them see the light, but rather because its opponents eventually die out and a new generation grows up that is familiar with it from the outset."[7] As a description of the actual course of the history of science, this assessment may be oversimplified, but as a cautionary remark for those who labor to introduce a new paradigm to the world, it implies a lesson. The only way members of the next generation can become familiar with a new theory is by learning about it. QBists, who believe that the personal experience of acquiring new information is the quintessential mechanism by which science evolves, are thereby counselled to promulgate their ideas widely and clearly. Broadcasting beats browbeating, according to Planck.

Chris Fuchs is the very embodiment of this strategy. With an engaging smile, a witty repartee, and inexhaustible enthusiasm, he roams the world like a latter-day troubadour. His lute is his laptop, his melody is mathematical, and his parchment is PowerPoint. Thus armed, he spreads the message of QBism across the globe. In the course of his travels, Chris has collected an amazingly wide-ranging circle of collaborators, colleagues, students, friends, and critics with whom he conducts a monumental e-mail correspondence. His aim is to make sure that even as the older generation of physicists (to which I belong), which was handed the conventional version of quantum mechanics by its predecessors, dies out, the new generation becomes familiar with QBism. Gradually, this effort is paying off as new converts are won over. I feel sure that in the end QBism will triumph as "a new scientific truth," a milestone along the long, winding road that began in the year 1900 with Max Planck's desperate quantum hypothesis.

Four Older Interpretations of Quantum Mechanics

Since the invention of quantum mechanics in 1925–1926, about a dozen distinct interpretations of the meaning of its mathematical formalism, each with numerous subvariants, have been proposed. Because none of them affect the practical applications of the theory, they are largely insulated from experimental corroboration or falsification. As a result few of them are ever completely withdrawn from the marketplace of ideas, though their relative popularities fluctuate. QBism is arguably the most radical interpretation. Instead of building on the accepted mathematical laws of quantum mechanics and adding theoretical superstructure to them, it digs down to their roots (*radix* is Latin for root) by revising the meaning of basic elements of the theory, such as probability, certainty, and measurement.

Here are four of the currently dominant interpretations in order of their popularity as measured by informal (and scientifically meaningless) polls of physicists.[1]

The Copenhagen Interpretation

This interpretation takes its name from Niels Bohr's institute in Copenhagen where the orthodox version of quantum mechanics was worked out, principally by Bohr and Heisenberg, with essential contributions from others. QBism retains many of the elements of the Copenhagen interpretation but disagrees fundamentally with some of them.

The observable properties of a quantum system are collectively called its *quantum state.* The quantum state in turn is described by a *wavefunction,* or equivalently, a matrix. In general, the wavefunction includes imaginary numbers such as the square root of -1. From the wavefunction, probabilities (real numbers between 0 and 1) are derived by standard rules. The probabilities refer to the possible outcomes of experimental observations and measurements.

A measurement somehow causes the instantaneous collapse of the initial quantum state to a new state corresponding to the actual outcome of the experiment. Repetitions of the experiment on the quantum system, which has been prepared in the same way for each trial, yield different outcomes in random order with different frequencies, like repetitions of the throw of a pair of dice.

While retaining the same mathematical formalism, QBism differs from Copenhagen in its interpretation of the wavefunction, the probabilities, and the collapse. The QBist wavefunction for a particular system is not a universally agreed-upon, observer-independent formula but an expression personal to each agent. It depends on each agent's knowledge and is thus subjective. QBist probabilities derived from the wavefunction are subjective Bayesian degrees of belief, rather than objective and frequentist. The collapse of the wavefunction is not a physical event—a change in the state of the system triggered by an experiment—but a Bayesian updating of a probability assignment upon the acquisition of new information.

The Many-Worlds Interpretation
The most direct way to avoid the problems of wavefunction collapse is to eliminate the collapse. This drastic move has garnered many adherents in recent years. The many-worlds interpretation assumes a single state of the universe with a wavefunction that evolves smoothly and predictably. In an experiment the wavefunction does not collapse. Instead, the entire universe, wavefunction and all, splits up into as many branches as there are possible outcomes. The observer is aware of only one of the outcomes and continues to live on that branch. Thus, the universe ramifies continuously into a vast multiverse in which every possible outcome actually occurs in one of a possibly infinite number of distinct, real universes that do not communicate with each other.

The principal objection to this interpretation is the exorbitant demand it makes on our imagination. More technical problems include its failure to account for the cause of the branching, and its difficulty in justifying the rules for deriving specific probabilities from the universal wavefunction.

The Pilot-Wave or Guiding-Field Interpretation

Inspired by the success of field theories such as electrodynamics and general relativity, several physicists, including Einstein for a while, favored an interpretation that starts with the accepted mathematical apparatus of quantum mechanics and rewrites it in a new format. This procedure yields an expression that resembles a real physical field of force that controls the motion of a particle in a deterministic, predictable way. This field is similar in kind but distinct from electromagnetic and gravitational fields. The suggestive image of a "quantum force" breaks down when several particles, say, N of them, are involved. The field in that case does not exist in our familiar three-dimensional space but in an abstract $3N$ dimensional space. While this unfamiliar property is shared by the conventional Copenhagen wavefunction, it detracts from the intuitive appeal of a guiding field. More troubling is the fact that the guiding field is explicitly nonlocal, like Newton's gravity. Amendments to the pilot-wave interpretation designed to render it compatible with special relativity and to include spin continue to be proposed and debated.

Spontaneous Collapse Theories

Because models of this kind add a completely new mechanism to the conventional quantum formalism, they should be called theories rather than interpretations. Collapses, in this view, are natural events that need no observer-induced triggers. They happen spontaneously but so rarely that they don't affect the interaction of individual small quantum systems. However, when a quantum system interacts with a large classical apparatus, such as a measuring instrument, the effect is amplified to the point that the entire wavefunction collapses. The disadvantage of this model is that the spontaneous collapse is an unexplained random event whose nature is as mysterious as the Copenhagen interpretation's observer-induced collapse it is designed to replace.

Notes

Introduction

1 George Gamow, *Mr. Tompkins in Paperback,* Canto Classics (Cambridge: Cambridge University Press, 2012).

1. How the Quantum Was Born

1 Helge Kragh, "Max Planck: The Reluctant Revolutionary," *Physics World,* December 1, 2000, 31–35, http://www.math.lsa.umich.edu /~krasny/math156_article_planck.pdf.

2 Frequency is measured in units of cycles per second, also called *hertz* and abbreviated Hz.

3 Since frequency has the dimensions of per second or inverse seconds, multiplying h by f cancels the seconds, leaving the quantum e with the metric unit of energy, the joule.

4 Phillip Frank, *Einstein—His Life and Times* (New York: Alfred A. Knopf, 1947), 71.

2. Particles of Light

1 "Do it Yourself Double Slit Experiment (Young's)—Easy At-Home Science," YouTube video, http://www.youtube.com/watch?v =kKdaRJ3vAmA.

4. The Wavefunction

1 $F = GmM/r^2$, where F stands for the strength of the gravitational force, G for the universal gravitational constant, m and M for the two masses that attract each other, and r for the distance between them.

2 A tiny tuning fork, as long as the width of a human hair and capable of displaying quantum behavior, was labeled "Breakthrough of the Year 2010" by the journal *Science.* See http://en.wikipedia.org/wiki /Quantum_machine.

5. "The Most Beautiful Experiment in Physics"

1 Unlike Max Planck, whose oscillator energy levels were off by a little bit, Niels Bohr had wrung the correct mathematical expression for the hydrogen energy levels out of his primitive mechanical model a dozen years before the invention of the wavefunction.

2 $F = ma$, where m is the mass of an object, a is its acceleration, and F is the net external force causing the acceleration.

3 The mathematical description of a wave usually includes positive and negative values to represent wave height above or below the x axis. But probabilities are never negative: they are real numbers between and including 0 and 1. Worse, wavefunctions usually include imaginary quantities, such as the square root of –1. So the numerical value of a wavefunction can't be *equal* to a probability. The mathematically correct recipe is this: "The probability density is equal to the wavefunction multiplied by its complex conjugate." I will simplify this by locutions such as: The wavefunction "yields" probabilities.

4 The experiment, including a video of the result, is described in "Feynman's Double-Slit Experiment Gets a Makeover," *Physicsworld.com,* March 14, 2013, http://physicsworld.com/cws/article /news/2013/mar/14/feynmans-double-slit-experiment-gets-a -makeover.

6. Then a Miracle Occurs

1 Isaac Newton to Richard Bentley, *Letters to Bentley, 1692/3,* third letter to Bentley, February 25, 1693, quoted in *The Works of Richard Bentley,* ed. A. Dyce, vol. 3 (London, 1838; repr., New York: AMS Press, 1966), 212–213.

7. Quantum Uncertainty

1 Wavelength $\approx xd/L$, where x is the distance between interference stripes, d is the distance between the two slits, and L is the distance from the slits to the screen.

2 Bram Gaasbeek, "Demystifying the Delayed Choice Experiments," July 22, 2010, http:www.arxiv.org/abs/1007.3977.

8. The Simplest Wavefunction

1 The amount of rotation of an ordinary object is measured by its angular momentum, which in turn depends on its mass, shape, and rotational velocity. Remarkably, the units of angular momentum turn out to be the same as those of Planck's constant h, a coincidence that helped to inspire the old Bohr model of the hydrogen atom.

2 "Raffiniert ist der Herr Gott, aber boshaft ist Er nicht," Alice Calaprice, *The Expanded Quotable Einstein* (Princeton, NJ: Princeton University Press, 2000), 241.

9. Troubles with Probability

1 D. M. Appleby, "Probabilities Are Single-Case, or Nothing," *Optics and Spectroscopy* 99 (2005): 447–462, http://arxiv.org/abs/quant-ph/0408058.

10. Probability according to the Reverend Bayes

1 The possessive form Bayes' is a compromise between Bayes's and Bayes.

2 See, for example, W. T. Eadie, D. Drijard, F. E. James, M. Roos, and B. Sadoulet, *Statistical Methods in Experimental Physics* (Geneva, Switzerland: CERN, 1971).

3 There is one important caveat: if the prior is exactly 0 or 1, no new information will budge it.

11. QBism Made Explicit

1 Carlton M. Caves, Christopher A. Fuchs, and Rüdiger Schack, "Quantum Probabilities as Bayesian Probabilities," *Physical Review* A 65 (2002): 022305–022315.

2 N. David Mermin, "Is the Moon There When Nobody Looks? Reality and the Quantum Theory," *Physics Today*, April 1985, 38.

12. QBism Saves Schrödinger's Cat

1 The quip is a paraphrase of a garbled version of a remark usually misattributed to different Nazi officials talking about *culture* rather than a *cat*.

13. The Roots of QBism

1 Quoted in Erwin Schrödinger, *Nature and the Greeks* and *Science and Humanism* (Cambridge: Cambridge University Press, 1996), 89.

2 Christopher A. Fuchs, N. David Mermin, and Rüdiger Schack, "An Introduction to QBism with an Application to the Locality of Quantum Mechanics," *American Journal of Physics* 82, no. 8 (2014): 749.

3 Werner Heisenberg, "The Representation of Nature in Contemporary Physics," *Daedalus* 87 (1958): 99.

4 Fuchs, Mermin, and Schack, "Introduction to QBism," 757.

5 N. David Mermin, "Quantum Mechanics: Fixing the Shifty Split," *Physics Today*, July 2012, 8.

14. Quantum Weirdness in the Laboratory

1 In 1964 John Bell raised the possibility of making the EPR thought experiments real. Laboratory implementations of his proposal began in the early 1980s and continue to this day.

2 Arthur Fine, "The Einstein-Podolsky-Rosen Argument in Quantum Theory," *The Stanford Encyclopedia of Philosophy*, Winter 2014, http://plato.stanford.edu/archives/win2014/entries/qt-epr/.

15. All Physics Is Local

1 Arthur Fine, "The Einstein-Podolsky-Rosen Argument in Quantum Theory," *The Stanford Encyclopedia of Philosophy*, Winter 2014, http://plato.stanford.edu/archives/win2014/entries/qt-epr/.
2 Christopher A. Fuchs, N. David Mermin, and Rüdiger Schack, "An Introduction to QBism with an Application to the Locality of Quantum Mechanics," *American Journal of Physics* 82, no. 8 (2014): 749–754.

16. Belief and Certainty

1 Arthur Fine, "The Einstein-Podolsky-Rosen Argument in Quantum Theory," *The Stanford Encyclopedia of Philosophy*, Winter 2014, http://plato.stanford.edu/archives/win2014/entries/qt-epr/.
2 In contrast to scientific and philosophical arguments by induction, mathematical proofs by induction are valid.
3 Christopher A. Fuchs, N. David Mermin, and Rüdiger Schack, "An Introduction to QBism with an Application to the Locality of Quantum Mechanics," *American Journal of Physics* 82, no. 8 (2014): 755.

17. Physics and Human Experience

1 Christopher A. Fuchs, N. David Mermin, and Rüdiger Schack, "An Introduction to QBism with an Application to the Locality of Quantum Mechanics," *American Journal of Physics* 82, no. 8 (2014): 749.
2 N. David Mermin, "QBism Puts the Scientist Back into Science," *Nature* 507 (March 27, 2014): 421–423.

18. Nature's Laws

The epigraph is taken from Max Planck, *Where Is Science Going?* trans. James Murphy (New York: W. W. Norton & Company, 1932), 107.

19. The Rock Kicks Back

1 Christopher A. Fuchs, "QBism, the Perimeter of Quantum Bayesianism," March 26, 2010, http://arxiv.org/abs/1003.5209.

2 Ibid.

3 Christopher A. Fuchs, "The Anti-Viejo Interpretation of Quantum Mechanics," April 25, 2002, 11, http://arxiv.org/abs/quant-ph /0204146. The article appeared before the word *QBism* was coined.

20. The Problem of the Now

1 N. David Mermin, "QBism as CBism: Solving the Problem of 'the Now,'" http://arXiv.org:1312.7825.

2 Rodolfo R. Llina's and Sisir Roy, "The 'Prediction Imperative' as the Basis for Self-Awareness," *Philosophical Transactions of the Royal Society* 364 (2009): 1301–1307.

21. A Perfect Map?

1 Lewis Carroll, *Sylvie and Bruno Concluded,* (London: Macmillan, 1893), chap. 11.

2 Marcus Appleby, "Concerning Dice and Divinity," November 26, 2006, http://arxiv.org/abs/quant-ph/0611261.

3 Ibid.

22. The Road Ahead

1 Richard Feynman, http://www.nobelprize.org/nobel_prizes /physics/laureates/1965/feynman-lecture.html.

2 Albert Einstein, "Über einen die Erzeugung und Verwandlung des Lichtes betreffenden heuristischen Gesichtspunkt," *Annalen der Physik* 17, no. 6 (1905): 132–148.

3 N. Bent, H. Qassim, A. A. Tahir, D. Sych, G. Leuchs, L. L. Sánchez-Soto, E. Karimi, and R. W. Boyd, "Experimental Realization of Quantum Tomography of Photonic Qudits via Symmetric Informationally Complete Positive Operator-Valued Measures," *Physical Review* X 5 (October 12, 2015): 1–12, http://journals.aps .org/prx/abstract/10.1103/PhysRevX.5.041006.

4 Rüdiger Schack, https://intelligence.org/2014/04/29/ruediger -schack/.

5 Jon Yard, http://physik.univie.ac.at/uploads/media/Yard_Jon_05 .06.14.pdf.

6 Ibid.

7 http://www.gutzitiert.de/zitat_autor_max_planck_thema _wissenschaft_zitat_27498.html.

Appendix

1 Adapted from Hans C. von Baeyer, "Quantum Weirdness? It's All in Your Mind," *Scientific American* 308, no. 6 (2013): 47.

Acknowledgments

My thanks go first and foremost to Chris Fuchs, who taught me all I know about QBism and thus unwittingly inspired this book. Conversations and correspondence with Marcus Appleby and David Mermin clarified many subtleties—I hope that I have presented their views fairly. Readers who patiently commented on successive versions of the manuscript included Roy Champion, Deke Dusinberre, Arthur Eisenkraft, Don Lemons, Tom Prewitt, and my brother Carl von Baeyer. Without the encouragement and support of my wife, Barbara Watkinson, and our daughter Madelynn von Baeyer, this book would never have been finished. My collaboration with our second daughter, Lili von Baeyer, the illustrator of this book, is always a pleasure. To all of them, I extend my heartfelt gratitude.

Index

Absolute space, 147

Absolute time, 147, 214

Abstraction, 81

Action at a distance, 67–71; QBism and, 168–169, 170; spooky, 158–159, 168–169

Agent: Bayesian probability and, 115–117, 188–189; measurement and, 205–206; uniqueness of Now to each, 215–216. *See also* Personal experience

Alice in Wonderland (Carroll), 219

Angular momentum, 242n1

Annie Oakley randomness, 55–56

Appleby, Marcus, 108–109, 112, 221–222, 230–232

Argument by induction, 178–179, 244n2

Argumentum ad absurdum, 177

Argumentum ad lapidem, 177

Aristotle, 57

Atom: Bohr model of, 36–40, 241n1; Schrödinger's cat paradox and state of, 141–142, 143

Atomic energies, 44–45

Atomism/atomists, 19–20, 23, 41; QBism and, 144–146

Bayes, Thomas, 91, 113

Bayesian probability, 5, 113, 114–127; certainty and, 179–182; degree of belief and, 115–119, 131, 188–189; example, 120–126; quantum mechanics and, 131–137

Bayes' law, 113, 119–120; certainty and, 179–180; equation, 122, 125–126; human motor control and, 218

Belief: Bayesian probability and degrees of, 115–119, 131, 188–189; certainty and, 184

Bell, John, 153, 190, 243n1

Berkeley, George, 136, 137

Betting, in Bayesian probability theory, 116–117

Binary digit, 90

bin Laden, Osama, 119

bit, 143; it from, 203–204

Black hole, 202

Blue light, 9

Bohr, Niels, 41; atomic model, 36–40, 82, 241n1; Copenhagen interpretation and, 236; on a priori given, 187–188, 189; on purpose of quantum mechanics, 149

Bohr radius, 38

Bose, Satyendra Nath, 102

Bose-Einstein statistics, 102, 103

Bullet, flight of, 53–54, 55–56, 63

Butterfly effect, 206

"Can the Quantum-Mechanical Description of Physical Reality Be Considered Complete?" (EPR), 156–161, 175, 177, 178–179

Carnap, Rudolf, 214

Carroll, Lewis, 219, 222

Cause and effect, law of, 57

Certainty: Bayesian probability and, 179–182; belief and, 184; nature's laws and, 200–201; prediction and, 220–221; QBism and, 181–184

Change, Bayesian probability and possibility of, 119–120

Classical physics, 1; Bohr atomic model and, 38–39; randomness and, 55–56

Climate science, Bayesian probability and, 126–127

Coin toss: Bayesian probability and, 117; frequentist probability and, 104–105, 106, 107–112, 227

Collapse of the wavefunction, 54–55, 63–66; older interpretations of quantum mechanics and, 237; QBism and, 132–135; *qubit* and, 82, 93–94; as scientific orthodoxy, 71–72; spontaneous collapse theories and, 239

Consciousness, QBism and, 221–222

Constructive interference, 27; quantum systems and, 43–44

Context, as determinative of Now, 217

Copenhagen interpretation, 236–237

Cromwell, Oliver, 180–181, 184

Cromwell's rule, 180–184, 201

Cube factory, 99–100

Dark energy, 39

Dark matter, 39

Dark spots, 26–27

Data compression, 200, 203–204

Degree of belief, Bayesian probability and, 115–119, 131, 188–189

Delayed choice experiment, 79–80

Democritus, 144–147, 149, 155, 204, 231

Destructive interference: electrons and, 33–34; light waves and, 26, 27–28; quantum systems and, 44

Dice throws, 11–12

Direction: spin, 86–87, 162–168; transitivity and, 162

Discreteness, 82, 93

Double-slit experiment: with electrons, 33, 59–62, 87, 151–152, 227–228; with photons, 27–28, 31–32, 59, 79–80, 140, 151–152; spin and, 87; uncertainty principle and, 77–80

Duration, frequency and, 76–77

Einstein, Albert: action at a distance and, 69; Bose's calculation and, 102; EPR paper, 156–161, 175, 177, 178–179; EPR paradox and, 156–158; on fusion of wave and particle theories, 34–35, 36; on God, 88–89; gravity and, 69, 70–71; guiding-field interpretation and, 238; icon of quantum mechanics and, 18; on the Now, 214, 216, 218; on objective reality, 136; perfect map and, 221; photo-electric effect and, 21–26, 30; photons and, 19, 20, 24–25, 225; quantum randomness and, 56, 57; special theory of relativity, 21, 147–148; spooky action at a distance and, 158, 159; theories of relativity and,

18, 21, 147–148, 170; thought process, 21

Electron gun, 54–55, 57, 59

Electrons, 20, 82–83; description of, 83; double-slit experiment, 33, 59–62, 87, 151–152, 227–228; exclusion principle, 102–103; Feynman diagrams, 171–174; GHZ experiment and, 161, 162–169; likened to platypus, 35–36; magnetism of, 83–84, 86, 88; mathematical equations predicting behavior of, 43; in photoelectric effect, 21–26; size of, 84; spin and, 83–90; wavefunction of, 63–65, 87–88; wave/particle duality and, 33–35, 36, 53–62

Elementary particles, sorting, 99–103

Energy: atomic, 44–45; harmonic oscillators and, 13–18, 46–47; photons and, 19

Energy density, frequency and, 9–11, 12–15

Entangled state, 162

Entanglement, Schrödinger's cat paradox and, 140, 143

EPR paradox, 157–158; experiments on, 161, 162–169, 243n1

Essential randomness, 56–57

Euclid, 227

Exclusion principle, 102–103

Experiments: delayed choice, 79–80; GHZ, 161, 162–169; as interaction between quantum systems, 207; quantum measurement and, 209, 210; unperformed, 142–143, 168, 169, 205

Fact creation, 206, 207–208

Fairness, frequentist probability and, 105, 107–112

Feldschlösschen, 211–213

Fermat, Pierre de, 231

Feynman, Richard: atomist manifesto, 144–145; double-slit experiment and, 59–62, 227–228; Feynman diagrams, 171–175; on hypothesis/guess, 196; reformulation and, 223, 224–225; on spin, 87; on understanding quantum mechanics, 2; Wheeler and, 202

Feynman diagrams, 171–175

The Feynman Lectures on Physics (Feynman), 59, 144–145

Free will, 201, 231

Frequency: duration of waves and, 76–77; energy density and, 9–11, 12–15; harmonic oscillators and, 13–18; of light waves, 9–11, 12–13; measurement of, 241n2; of sound waves, 13–18, 45

Frequentist probability, 5, 104–112, 113–114, 115, 118, 127

Fuchs, Christopher (Chris): locality and, 175, 176; QBism and, 2–3, 4, 112, 233; on quantum dimension, 230; on quantum experiments, 209, 210; on role of agent in QBism, 189, 205–206, 208; Schrödinger's cat and, 138; Wheeler and, 202, 203

Fundamental theorem of arithmetic, 227

Galileo, 160

Gambler's fallacy, 106, 107–108, 118

Gamow, George, 1

General theory of relativity, 70, 151, 170

GHZ experiment, 162–169, 170, 176

GHZ rule, 164, 166–167

Gluons, 173

God: knowing mind of, 220, 221–222; laws of nature and, 197–198

Going to the limit, 182

Gravity: Einstein's theory of, 70–71; universal gravitation, 42, 66–67, 68–70, 170, 197, 198, 241n1

Greenberger, Daniel, 162

Guiding-field (pilot-wave) interpretation, 238–239

Harmonic oscillator, 13–18; energy levels, 241n1; uncertainty principle and, 74; wavefunction of, 46–47, 50, 52

Harris, Sidney, 65

Hawking, Stephen, 138

Heisenberg, Werner, 73–76; Copenhagen interpretation and, 236; on objective reality, 149–150

Heisenberg cut, 152–155

Heisenberg's microscope, 75–76, 80

Hertz, 241n2

Heuristic role, QBism and, 225–233

Hidden variables, 134–135; GHZ experiment and, 168–169

Higgs boson, 24, 173

Hilbert, David, 229

Horne, Michael, 162

Human experience, physics and, 187–195

Hydrogen, atomic model, 36–38

Hypothesis, transition to law, 196–197

Induction, argument by, 178–179, 244n2

Inductive reasoning, 196

Information: Bayesian probability and new, 119–127; as key to understanding nature, 203–204

Infrared frequencies, 10

Intellect, perceiving nature through, 146–147, 155

Interference effects, 26, 27–30, 140; constructive, 27, 43–44; destructive, 26, 27–28, 33–34, 44; displays of, 28–30; quantum systems and, 43–44

Intrinsic randomness, 56–56

Johnson, Samuel, 177, 187, 209

Jordan, Ernst Pascual, 136

Joule, 241n3

Korzybski, Alfred, 50, 91

Laplace, Pierre-Simon, 113

Law of cause and effect, 57

Law of total probability, 227

Laws of nature. *See* Nature's laws

Light: nature of, 23–24; as particles, 25–26, 31–40. *See also* Photons

Light waves, 9–11, 12–13, 22–23, 26; double-slit interference and, 27–28, 31–32; interference effects and, 26, 27–30

Lindley, Dennis, 180–181

Local action, 70–71

Locality, 158–159, 160–161, 171–175; GHZ experiment

and, 167–169; QBism and, 175–176; sharing the Now and, 216

Magnetism, of electron, 83–84, 86, 88
Magnitude, of spin, 85–86, 88
Many-worlds interpretation, 237–238
Map: ideal, 219–222; wavefunction compared to, 47–48, 50–51
Mathematical abstractions, in physics, 81
Mathematical form, of wavefunction, 44
Mathematical models, 41–43; harmonic oscillator, 46–47, 50, 52; perfect, as goal of physical science, 219
Matrix (matrices), 50, 90, 236
Meaning: of probability, 127; of quantum mechanics, 187–188, 224–225
Measurement: in Copenhagen interpretation, 236; of frequency, 241n2; quantum mechanics and, 204–205, 209, 236; quantum uncertainty and, 75–76; of spins, 164
Mechanical models, 41, 42–43
Mermin, N. David, 153–154, 175, 176, 194, 215–216
Metaphysics, Wheeler and, 203
Miracle cartoon (Harris), 65
Motion: quantum mechanical law of, 63–64, 66; universal law of, 63, 64
Musical instruments, frequencies in, 44–45

Nature, ways of perceiving, 145–155. *See also* Reality

Nature (journal), 194
Nature's laws, 68; invention of, 196–198, 199; status of, 198–201
Neutrinos, 173
Newton, Isaac: absolute space/absolute time and, 147; action at a distance and, 68–69; laws of nature and, 197; objective reality and, 151; time and, 147, 214; universal law of gravitation, 42, 66–67, 68–70, 170; universal law of motion, 63, 64
Newtonian physics. *See* Classical physics
Now, the problem of the, 211–218

Oakley, Annie, 53–54
Obama, Barack, 118–119
Objective, QBism and subjective vs., 154
Objectivity, frequentist probability and, 104–105
Observation, 196; stopping flow of time and, 211–213
Observer, 204, 207
Observer effect, uncertainty principle and, 75–76, 78–80
One (1), Bayesian probability and, 117, 121, 180–183

Paradoxes: EPR, 156–158, 161, 162–169; Schrödinger's cat, 2, 138–143, 154–155; Wigner's friend, 135–137, 138, 187, 192
Parameter, 16
Participatory universe, 204–210
Particle physics, standard model of, 173–174
Particles, velocity and positions of, 73. *See also* Wave–particle duality

Particle statistics, 101–103

Particle theory of light, 25–26

Peres, Asher, 142, 143, 168, 169

Personal experience: models of the world and, 187; QBism and, 175–176, 188–195. *See also* Agent

Photoelectric effect, 21–26, 30

Photons, 19, 20, 24–25, 82–83; Bose-Einstein statistics and, 101–102; delayed choice experiment, 79–80; double-slit experiment, 27–28, 31–32, 59, 79–80, 140, 151–152; Feynman diagrams and, 171–173; indistinguishable quality of, 101; wave/particle duality and, 31–32, 35, 36. *See also* Light

Physics: division of modern, 151; goal of, 41; psychology and, 231–232. *See also* Classical physics; QBism; Quantum mechanics

Physics World (journal), 59

Pilot-wave (guiding-field) interpretation, 238–239

Planck, Max, 41, 203, 233; creation of quantum and, 9–20; harmonic oscillators and, 14–18, 241n1; on nature's laws, 196, 197, 198, 201; radiation curves and, 11–13, 16–18, 23–24; radiation law, 102; on scientific truths, 232

Planck-Einstein equation $(e = hf)$, 16–18, 23, 36–37, 45, 47, 77

Planck's constant, 17–19, 46, 77, 228, 230, 242n1; Bohr atomic model and, 36; spin and, 85–86

Platypus, electrons likened to, 35–36

Podolsky, Boris, 156–158, 175, 177, 178–179

Point particle, rotation and, 84–85

Position of particle, 73

Posterior probability, 125

Prediction: certainty in, 220–221; influence on the Now, 217–218

Present, time and, 213–214

Prior probability, 125

Probability, 5, 97–104; Bayesian (*see* Bayesian probability); cube factory, 99–100; electron gun pattern and, 58–59; formula, 98; frequentist, 5, 104–112, 113–114, 115, 118, 127; law of total, 226; QBism and, 225–228; quantum law of, 228–230; single-case, 106–107, 109–112; superposition and, 92, 94; wavefunction and, 59–62, 82, 242n3; wavefunction collapse and, 65–66

Probability theory, 98–99

Proton, 38

Psychology, physics and, 231–232

QBism (Quantum Bayesianism), 3–4, 6; avoiding spooky action at a distance and, 168–169; Bayesian probability and, 113, 114, 118; certainty and, 181–184; collapse of the wavefunction and, 132–135; consciousness and, 221–222; Copenhagen interpretation vs., 236–237; creation of, 131–132; Cromwell's rule and, 181–184; Heisenberg cut and, 153–154; heuristic role of, 225–233;

intrinsic randomness and, 57; locality and, 175–176; meaning of quantum mechanic concepts and, 224–225; mind of God and, 220, 221–222; paradox of Wigner's friend and, 135–137, 138; participatory universe and, 204–210; probability 1 and 0 assignments, 181–183; probability as degree of belief of single agent, 188–189; problem of the Now and, 215–218; *qubit* vs., 90–91; as radical interpretation of quantum mechanics, 235; reality and, 177, 208–209; roots of, 144–155; Schröding-er's cat paradox and, 141–143, 154–155; status of nature's laws in, 199–201; thesis of, 131; value of personal experience and, 188–195; wavefunction and, 237

Quantum: creation of, 9–20; joule and, 241n3; why the, 202–203, 224

Quantum Bayesianism. *See* QBism (Quantum Bayesianism)

Quantum Bayesianism (Fuchs), 3

Quantum dimension, 229–230

Quantum effects, system size and, 151–152

Quantum electrodynamics (QED), 171, 223

Quantum force, 238

Quantum information theory, 2, 4

Quantum law of total probability, 228–230

Quantum mechanics, 1–4, 5–6; applications, 150–151;

Bayesian probability and, 131–137; Copenhagen interpretation, 236–237; double-slit experiment and, 59–62; EPR paradox and, 157–158; interpretation of Schrödinger's cat paradox, 139–141, 142–143; law of motion, 63–64, 66; locality and, 158–159; many-worlds interpretation, 237–238; mathematical models and, 42–43; meaning of, 187–188, 224–225; measurement and, 204–205, 209; objective reality and, 148–155; observer and, 204; older interpreta-tions of, 235–239; pilot-wave (guiding-field) interpretation, 238–239; QBism and meaning of concepts of, 224–225; realism and, 159–161; sponta-neous collapse theories, 239; suspicion of philosophy, 103; wavefunction and, 45–46. *See also* QBism (Quantum Bayesianism)

"Quantum Probabilities as Bayesian Probabilities" (Caves et al.), 132

Quantum randomness, 56–57, 60

Quantum state, 236

Quantum weirdness, 2, 3, 6, 151; EPR paradox experiments, 161, 162–169; Schrödinger's cat and, 141; thought experiments, 156–158

Quarks, 173

qubit (quantum bit), 90–94, 143; experimental use of, 161; GHZ experiment and, 161, 162–169; quantum dimension of, 229

qubit wavefunction, 91–94;
paradox of Wigner's friend
and, 135–137

Radiation curves, 10–13, 16–18,
23–24
Randomness, 55; Annie Oakley,
55–56; quantum (essential,
intrinsic), 56–57, 60
Realism: EPR and, 159–161; GHZ
experiment and, 165–169;
locality and, 175
Reality: *bit* as key to understanding, 203–204; Bohr on,
187–188; EPR paradox and,
157–161; perception of,
145–155, 177–179; QBism and,
136–137, 208–209; quantum
mechanics as technique for
comprehending, 192; quantum
theory and, 148–155; special
theory of relativity and
perception of, 147–148; wave/
particle duality and perception of, 148–149
Really Big Questions (RBQs),
Wheeler and, 202–203
Red light, 9–10
Reformulations, 223, 224–225
Relative time, 214
Relativity theory, 18; general, 70,
151, 170. *See also* Special
theory of relativity
Rosen, Nathan, 156–158, 175, 177,
178–179
Rotation: angular momentum and,
242n1; of point particle, 84–85
Roulette parable, 108–111
Rule, 200

Schack, Rüdiger, 175, 176, 232
Schrödinger, Erwin: on object-
subject relations, 150;

Schrödinger's cat and,
139–141; wavefunction and,
43–44, 45, 51
Schrödinger's cat paradox, 2,
138–143; QBism and, 154–155
Scientific method, 193–194
Scientific realism. *See* Realism
Scientific truths, 232–233
Senses, perceiving nature
through, 146–147, 149, 155
Shared experiences, QBism and,
191–192
Signal, 79–80
Single-case probability, 106–107,
109–112
Soap bubbles, interference
effects and, 28–29
Sound waves, 44–45, 77. *See also*
Harmonic oscillator
Special theory of relativity, 21,
151; action at a distance and,
70, 170; Bohr model and, 37;
break with absolute
objectivity and, 147–148; laws
of, 198; pilot-wave interpreta-
tion and, 238–239
Spin, electrons and, 83–90; GHZ
experiment and, 162–169;
transitivity and, 162
Spin wavefunction, 88–90
Spontaneous collapse theories,
239
Spooky action at a distance,
158–159; QBism and,
168–169
Spooky effect, 167
Standard model of particle
physics, 173–174
*Stanford Encyclopedia of
Philosophy*, 159
Statistical randomness, 55–56
Subjective, QBism and objective
vs., 154

Superposition: of light waves, 26; quantum systems and, 43–44; *qubits* and, 92; of wavefunction, 47, 82, 226

Thomson, G. P., 33
Thomson, J. J., 33–34, 35
Thought experiments, 156–158, 243n1
Time: absolute, 147, 214; the present and, 213–214; relative, 214; stopping flow of, 211–213
Transitivity, 161–162, 164–165
Tuning fork, 14, 52, 152, 241n2

Ultraviolet light, 10
Uncertainty principle, 73–81; observer effects and, 75–76, 78–80; spin and, 87
Universal gravitation, 42, 66–67, 68–70, 170, 197, 198, 241n1
Universe: in many-worlds interpretation, 237; participatory, 204–210
Unperformed experiments, outcomes and, 142–143, 168, 169, 205

Van Fraassen, Bas, 99
Velocity of particle, 73
Venn, John, 113

Water waves, 32; duration and frequency of, 76–77
Wavefunction, 41–51; Copenhagen interpretation and, 236; of electron, 54–55, 87–88; many-worlds interpretation and, 237; map

analogy, 47–48, 50–51; numerical value of, 242n3; paradox of Wigner's friend and *qubit,* 135–137; probability and, 59–62; properties of, 82; QBism and, 137, 225–226, 237; *qubit,* 91–94, 135–137; spin, 88–90; uncertainty principle and, 73–81; wave/particle duality and, 52–53. *See also* Collapse of the wavefunction
Wavelength: formula, 242n1; which-path information and, 77–80
Wave/particle duality, 31–40; Bohr atom and, 38–39; of electrons, 53–62; objective reality and, 148–149; wavefunction and, 52–53
Waves: sound, 44–45, 77; water, 32, 76–77. *See also* Light waves
"Wavicle," 35
Wheeler, John Archibald, 202–204, 206, 207–208, 224
Wigner, Eugene, 135–137
Wigner's friend paradox, 135–137, 138, 187, 192
Winter, Rolf, 35–36
Witness detector, 79–80
Worldview, QBist, 192–195

Yard, Jon, 229
Yellow light, 9
Young, Thomas, 26, 80, 140

Zeilinger, Anton, 162
Zero (0), Bayesian probability and, 117, 121, 179–183